T0360545

Ecology of the Acanthocephala

Commonly known as spiny-headed worms, acanthocephalans have
a characteristic eversible proboscis, which bears rows of recurved
spines used for attachment to the intestine of their host.
Acanthocephalans are endoparasitic and an extremely successful
group found in almost all marine, freshwater and terrestrial systems,
infecting a huge range of definitive (usually vertebrate) and
intermediate (usually arthropod) hosts during their life cycles. This
volume is concerned with the ecology (distribution and abundance)
of the Acanthocephala, and through this approach aims to reveal the
huge success of this group of parasites. The acanthocephalans have
evolved differently from all other groups of parasites, and as such
represent a distinct and alternative pathway of parasite evolution
and host–parasite interactions. Written for parasitologists, ecologists
and zoologists who are interested in learning about a different
pathway of parasite evolution, this book is aimed at postgraduate
and research scientists.

C. R. KENNEDY is Emeritus Professor of Parasitology at the
University of Exeter.

Ecology of the Acanthocephala

C. R. KENNEDY
School of Biological Sciences,
University of Exeter

CAMBRIDGE
UNIVERSITY PRESS

Shaftesbury Road, Cambridge CB2 8EA, United Kingdom

One Liberty Plaza, 20th Floor, New York, NY 10006, USA

477 Williamstown Road, Port Melbourne, VIC 3207, Australia

314–321, 3rd Floor, Plot 3, Splendor Forum, Jasola District Centre, New Delhi – 110025, India

103 Penang Road, #05–06/07, Visioncrest Commercial, Singapore 238467

Cambridge University Press is part of Cambridge University Press & Assessment,
a department of the University of Cambridge.

We share the University's mission to contribute to society through the pursuit of
education, learning and research at the highest international levels of excellence.

www.cambridge.org
Information on this title: www.cambridge.org/9780521850087

First published 2006

A catalogue record for this publication is available from the British Library

ISBN   978-0-521-85008-7   Hardback

Contents

Preface

To do science is to search for repeated patterns, not simply to accumulate facts. The only rules of the scientific method are honest observations and accurate logic. To be great science it must also be guided by a judgement, almost an instinct, for what is worth studying.

MacArthur (1972)

We conclude that the complicated interplay between virulence and transmissibility of parasites leaves room for many coevolutionary paths to be followed, with many endpoints.

Anderson & May (1982)

The Acanthocephala are generally considered to be a small and rather insignificant group of parasites. They are a minor phylum of interest to specialists; as they rarely, and then accidentally, infect humans or cause significant disease or disorder to their farmed and domestic animals, they receive a cursory treatment even in most parasitology textbooks. Nevertheless, there have been some books devoted exclusively to them, principally *Acanthocephala of Domestic and Wild Animals* (Petrochenko, 1956, 1958), Vol. V of *Systema Helminthum* (Yamaguti, 1963), *An Ecological Approach to Acanthocephalan Physiology* (Crompton, 1970) and *Biology of the Acanthocephala* (Crompton & Nickol, 1985). The first two of these books are fundamentally systematic in their approach, whereas the latter two are far more biological. Crompton & Nickol's (1985) book does contain some chapters on acanthocephalan ecology, but it was published 20 years ago, and more recent information on their ecology can only be found in the primary scientific literature. Much recent information is being presented at the International Acanthocephalan Workshops, but proceedings of these are not published.

Parasitology can be viewed as a special branch of ecology since it is concerned with the relationships between organisms and their environment. This present volume attempts to present a comprehensive account of the ecology of the Acanthocephala. By so doing, it aims to show how an ecological approach can open up a new perspective on the phylum such that acanthocephalans can now be seen as a common, widespread and very successful group of parasites. The pattern of their evolution has differed in many respects from those of all other groups of helminth parasites and so they present a distinctive and characteristic path of host—parasite co-evolution.

Their impact on their hosts extends beyond host individuals onto populations and communities of free-living organisms and indeed into whole ecosystems. They are also proving to be very sensitive indicators of some forms of pollution. It is intended, therefore, that this book will be of interest to a wider readership, especially to freshwater and marine biologists and also to ecologists in general, as well as to parasitologists who may be excited by the prospect of recognising and studying a different pattern of parasite evolution. It is aimed primarily at postgraduates and researchers and it assumes at least an undergraduate knowledge of parasitology and ecology. It challenges some current ideas and paradigms on biodiversity, by showing what can be achieved in evolutionary terms by a group of organisms that comprises a small number of species and exhibits a high degree of morphological uniformity. It aims to provoke an interest in the group and to stimulate the reader. It also attempts to pluck the phylum from the obscurity of current textbooks and place it on a par with, for example, cestodes and nematodes as being a group worthy of study for its own sake.

No book can be comprehensive in its coverage, and this one is no exception. I am solely responsible for the selection of examples, which have been chosen from a wide range of hosts and habitats around the world. Nearly all examples in figures and tables have appeared in published form previously and I must thank all the authors and publishers for permitting me to use them: the source of each is acknowledged fully in the legends to figures and captions to tables. While the author alone must be held responsible for presentation and any errors, no book can ever be completed without the help of many people. I would like particularly to thank Ward Cooper, the then Commissioning Editor at Cambridge University Press, for his advice and efforts on my behalf and also his successor Dominic Lewis for being available to answer my many queries and deal with problems. Many of the ideas presented here have been developed in discussion with friends and colleagues over a long period of time; and although it is invidious to single out particular persons I would like to mention Al Bush, David Crompton, Bahram Dezfuli, Celia Holland, John Holmes, Brent Nickol and Phil Whitfield in particular who have all contributed, often unknowingly and unwittingly, to the development of my ideas. I owe them a great deal as colleagues and friends. Other ideas have emerged from discussions with my postgraduate students and postdoctoral fellows and to them, and especially to Rachel Bates and Alastair Lyndon, I express my thanks.

Many other people have helped me in a variety of ways, by providing me with photocopies and advice, and I would particularly like to express my thanks to Omar Amin, Chris Arme, David Gibson, Eileen Harris, Robert Poulin, Jamie Stevens and Bernd Sures. Two other people deserve special thanks. The first of these is Ian Tribe, who, although a botanist, read through the manuscript for me and made many very valuable suggestions. Finally, and above all, I must thank my wife Pat. My interest in acanthocephalans was set alight by some of the questions raised in her thesis and now 30 years later I am still trying to explain and understand the significance of some of her findings. She was in large measure responsible for planting the idea of this book in my mind and in getting me to start on it despite my insistence that I would never write another book. She has supported me and encouraged me continually throughout the writing, through the creative times and through the inevitable frustrating times. Quite simply, this book could not have existed without her and both I and anyone who reads it and enjoys it owes her an enormous debt.

1

Introduction

In biology, as in all natural sciences, particular concepts come into and go out of fashion. Biodiversity, including species richness, is currently in fashion. This is in large measure a consequence of concerns about species rarity and threats of global and local extinction of organisms in response to human activities and global warming. Conservation interests in particular stress the importance of biodiversity and the need to preserve or restore it as the case may be. The term is found in a wide range of conservational, ecological and biological literature. In practice it is seldom defined and even though it is clear that it means different things to different people, it is always regarded as 'a good thing'. Definitions aside, biodiversity in practice is often used to justify the conservation of a rare species or of a particular habitat on a local or a regional scale. Emphasis on rare species has always been a feature of research and conservation interests. This has often been at the expense of understanding widespread, common and successful species and the ways in which they can adapt to human influences and their consequent changes in habitat and land use.

Biodiversity concerns are also frequently subjective and anthropocentric. They are all too often centred on particular types of habitat, communities and ecosystems that contain species that are considered particularly attractive. There is more concern about the possible loss of one species of mammal or bird than ten species of insect or crustacean. Moreover, there is also a strong bias in conservation towards free-living organisms and communities and against organisms that may cause disease or be considered harmful in any other context. Concerns about the reductions in biodiversity stemming from loss of tropical rainforest or heathland, for example, are not matched by a comparable concern

1

about any loss of communities of bacteria or viruses; and protestors about the possible extinction of the smallpox virus are not thick on the ground, even though this would be a deliberate and wilful human act of extinction. Indeed, it is very seldom appreciated that superimposed on the populations and communities of free-living organisms there are populations and communities of parasitic ones. Since every free-living plant or animal is actually or potentially a host to one or several species of parasitic organism it is likely that there are more species of parasitic organisms than there are of free-living ones (Price, 1980). Parasitic populations and communities are not only superimposed on those of free-living animals but also may interact closely with them. Parasites are an integral and functional part of any ecosystem, and parasite diversity is thus an integral part of biodiversity.

Parasitology, however, is still all too often taught as an independent discipline with minimal or no attempt to integrate it with other biological disciplines such as ecology. Teaching of parasitology often still tends to focus primarily upon parasitic diseases of man and his domestic animals and crops and on the harmful effects of parasites on their hosts. Treatment in textbooks concentrates on malaria, bilharzia and eelworms, for example. Although this is set in the context of parasite species richness and diversity in structure, and especially diversity and complexity of parasitic life cycles, the emphasis is still on medical or veterinary examples. Despite the difficulties in remembering all the hosts and larval stages of parasites, students and the wider public seldom fail to be impressed by the frequently bizarre aspects of parasitic life cycles and by the success and impact of parasites.

Parasitology is nevertheless fundamentally an ecological discipline as it is concerned with the relationships between two species, the parasite and its host—with the relationships between an organism, the parasite, and its environment, which is another living organism, the host. Described thus, parasitology could be considered a specialist branch of ecology. Without attempting to define parasitism (most parasitology textbooks will normally provide one or several, often conflicting, definitions), I should point out that Russian parasitologists in particular, and especially V. A. Dogiel, have always emphasised the ecological nature of parasitism and could be considered to have actually founded the study of ecological parasitology. Dogiel (1964) himself stated clearly that parasitism is an ecological concept, and so parasitology should concern itself with those animals which use living animals of other species as their environment and source of food. Kennedy (1975) also emphasised the value of an ecological approach in understanding parasites and their relationships with their hosts. More recently, Combes (2001) has emphasised the prolonged nature of the host–parasite interactions such that the association of parasite– host can be viewed as a system, with novel characteristics of its own that are not just the simple sum of its components. His views are expressed most succinctly in the title of his book *Parasitism: The Ecology and Evolution of Intimate Interactions.* Since parasite–host systems are integral parts of every ecosystem, and since parasites can affect the life of practically every other organism (Price, 1980), one might expect the ecological literature to give as much prominence to parasites as to free-living organisms. However, this is not the case as very few ecology textbooks, with the notable and laudable exception of Townsend *et al.,* (2000), consider parasites at all: they are left to zoological, and specifically parasitological, texts. General zoological texts and parasitological texts tend to concentrate on the parasitic Protozoa, Platyhelminthes and Nematoda as these phyla contain the greatest number of parasitic species, including those pathological to man, his domestic animals and his crops. Other small parasitic groups, and parasitic members of phyla of predominantly free-living species, are ignored or at best given a cursory treatment.

Amongst these groups are the Acanthocephala, a small monophyletic phylum of which all species are obligatory endoparasites: it is in fact one of only two phyla to be exclusively parasitic and with no free-living members (the other is the Nematomorpha). Earlier zoological texts, such as Barnes (1963), devote only a few paragraphs to them, stressing the large numbers that may be found in a host and the damage this may do to the host intestine; and even in some later texts, for example Barnes *et al.* (1988), they are given a very cursory treatment. Many current general biology texts, however, for example Campbell (1996), do not even mention the phylum. Even parasitology texts give them little prominence in comparison with cestodes or digeneans. Cox (1993), for example, does not discuss them at all in the systematic section of the book, although he devotes seven pages to cestodes, and only mentions a few specific examples as appropriate. Smyth (1994), however, does devote a short chapter exclusively to the Acanthocephala, but stresses their lack of diversity in structure, life cycles and habits. Roberts & Janovy (1996) also devote a chapter to them, but claim (incorrectly) that they are seldom encountered and are rare in comparison with cestodes and nematodes, and that they are also capable of traumatic damage to their hosts, on occasion attaining epizootic levels leading to host mortalities. Only Bush *et al.* (2001)

give them extensive coverage and treat them on a par with cestodes. Most authors consider the Acanthocephala as a minor phylum, equating this with their being relatively unimportant. Even a book on parasite ecology (Kennedy, 1975) does not single them out for any special emphasis.

Until recently, the few books devoted exclusively to the Acanthocephala tended to be primarily or even exclusively systematic in treatment. Included within this category are Meyer (1938), Petrochenko (1956, 1958) and Yamaguti (1963). The most recent books are those of Crompton (1970), which adopted an ecological approach to acanthocephalan physiology, and Crompton & Nickol (1985), in which contributors covered many aspects of acanthocephalan biology and in which were chapters on epizootiology, life history models and population dynamics. Until now, however, there has been no single book that has been devoted exclusively to the ecology of the Acanthocephala.

This disproportionate treatment of the Acanthocephala and their ecology probably reflects three things: the relatively small number of species, their relative lack of pathogenicity to their vertebrate hosts and the perceived lack of diversity in acanthocephalan structure and life cycles. The Acanthocephala are undeniably only a small phylum with current estimates of around 1000+ species (based on Amin, 1985a). They may cause local damage to the intestine of their vertebrate hosts, but they are very seldom the cause of any serious damage or death to man himself or to his domestic animals. It is also undeniable that they are characterised by great uniformity of structure, larval stages and life cycles. They are in many ways a systematist's nightmare as they have few organs upon which to base their taxonomy (Brown, 1987). The only hard structures they possess are the hooks on their probosces (Fig. 1.1), and we do not even begin to understand the adaptive significance of the differences in hook numbers, size and arrangement between species. Moreover, there is considerable intraspecific variation in the arrangement of the hooks (Brown, 1987). The internal structures are few and remarkably similar (Fig. 1.2), differing between species only in such details as the number and arrangement of cement glands in the males or the position of the neural ganglion. Furthermore, all species have the same fundamental life cycle and developmental stages: all have a free-living egg (acanthor), all require an arthropod intermediate host for the larval acanthella and cystacanth stages and all utilise a vertebrate definitive host as adults (Fig. 1.3).

This lack of anatomical diversity may, however, be deceptive, as will become evident, and the rigidity of the life cycle may be more

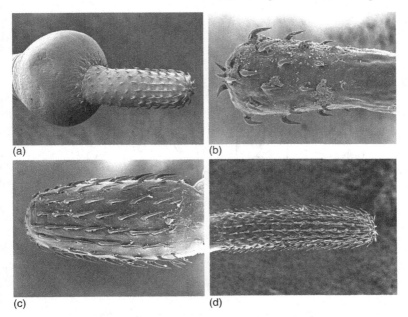

Fig. 1.1 Scanning electron micrographs of acanthocephalan probosces.
(a) *Pomphorhynchus laevis* from chub *Leuciscus cephalus*; (b) *Acanthocephalus anguillae* from chub; (c) *Acanthocephalus lucii* from perch *Perca fluviatilis*;
(d) *Acanthocephalus clavula* from eels *Anguilla anguilla*.

apparent than real. Concentration on such uniformity is certainly misleading as it deflects attention from the accomplishments of the acanthocephalans as parasites and from their ecological achievements. However success may be defined, they can be considered a highly successful group of parasites in that they infect all classes of vertebrates: they are to be found in the sea, in fresh water, on land and, in birds, in the air, and they occur on all continents and in all biomes. In comparison, it has taken the parasitic platyhelminths some 25 000 species and an enormous diversity of larval stages and life cycles to achieve a similar distribution across hosts and in space (Kennedy, 1993a). Throughout the phylum as a whole, the platyhelminths exhibit wide diversity in larval stages, especially within the cestodes, and in the number and identity of intermediate hosts utilised. They may be said to exhibit high diversity and high achievement, whereas the acanthocephalans have attained a similar high level of achievement with, apparently, minimal diversity. Even the parasitic Crustacea, with almost 3000 more species and which exhibit a more diverse use of hosts in their life cycles, have failed to make the transition to infecting

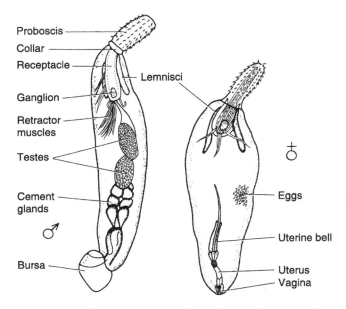

Fig. 1.2 Morphology of an adult male and female *Acanthocephalus* sp.
(From Smyth, 1994.)

terrestrial vertebrates. The nematodes, which are similar to acanthocephalans in some respects in that they exhibit a high degree of
uniformity of adult and larval structure and life cycle stages, show very
wide flexibility and diversity in numbers and identity of intermediate
hosts in their life cycles, in the stage at which they are parasitic
and also in the degree of pathogenicity towards their definitive hosts.
The acanthocephala are thus one of the smallest and least diverse
groups of metazoan parasites, yet are as widely distributed amongst
vertebrate hosts and biomes as are the larger and more diverse
parasitic groups.

Clearly there is something distinctive about the acanthocephalans: they appear to have evolved as a group and co-evolved with their
hosts in a different way from other parasitic groups. The aim of this
book is to challenge many of the assumptions about them that may
make them initially appear dull, uniform and uninteresting, even
to many parasitologists, by showing that such appearances may be
deceptive. It is the intention of this book to show that an ecological
approach can open up a whole new perspective on the acanthocephalans. In this light, the phylum appears very different. Such an approach
to the group is novel and differs from previous treatments of the

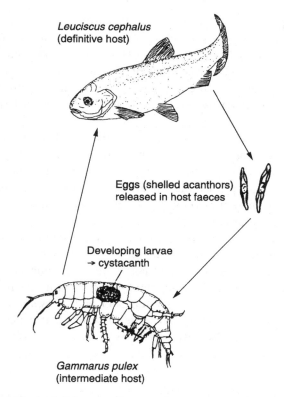

Leuciscus cephalus
(definitive host)

Eggs (shelled acanthors)
released in host faeces

Developing larvae
→ cystacanth

Gammarus pulex
(intermediate host)

Fig. 1.3 The basic life cycle of an acanthocephalan. *Pomphorhynchus laevis* eggs released into fresh water in the fish's faeces are eventually eaten by a *Gammarus*. They hatch in the intestine and the released acanthors move into the haemocoel, where they develop into the orange-coloured cystacanth. This infective stage remains in the *Gammarus* until the amphipod is eaten by the fish, when the larva develops into an adult in the intestine and the cycle is completed. (From Brown & Thompson, 1986, with permission, Institute of Biology.)

phylum. It involves a change in emphasis and a concentration on many aspects of acanthocephalan biology other than morphology. For example, it will be shown that the acanthocephalans can exhibit diversity at the molecular level and that they can escape the restriction of requiring only a single intermediate host in their life cycle. They can have a major impact on their intermediate hosts at individual and population levels by altering the behaviour of infected individuals which results in the host's death by predation. This in turn means that they can have important effects on the communities and food webs of free-living organisms. This has been demonstrated to be the case for virtually all freshwater species, yet even some of the very best freshwater textbooks such as that of Moss (1988) make no mention of their effects on individuals, populations and communities of freshwater organisms, any more than books on freshwater pollution such as that by Mason (1991) discuss their role as bioindicators of heavy metal pollution.

It is the contention of this book that the key to understanding the success of the Acanthocephala and their different pattern of evolution lies in an understanding of their ecology. One can then begin to appreciate that their relationships with their intermediate hosts may be more important in ecological and evolutionary contexts than their relationships with their vertebrate hosts. It may also help one to appreciate that the term minor phylum should not carry any implications of importance, as the impact of members on free-living organisms may be out of all proportion to the numbers of species. Above all, it may cause us to think harder about the importance of diversity: it is not the beginning and end of everything and it should be instructive to all biologists to realise that a phylum with apparently little biodiversity may be as successful as many far more diverse phyla.

1.2 AN ECOLOGICAL APPROACH TO THE ACANTHOCEPHALA

Ecology can be defined in many ways. At its simplest it can be defined as the study of the relationships between an organism and its environment, and this of course is effectively a truism as far as parasites are concerned since their host is their environment. Russian parasitologists (Pavlovski, 1934, and Dogiel, 1964, in particular) further distinguish two types of environment in the case of parasites: the micro-environment, or the immediate environment within the host, and the macro-environment, or the external environment of the host. However, in the context of the present book it is the definition of Andrewartha & Birch (1954) that has been adopted: ecology, in their view, is the study of the factors affecting the distribution and abundance of an organism (or, strictly, the abundance, since absence of a species is the same as zero abundance). They considered these factors under headings such as food, a place to live and impact of other species; but this is not the easiest approach for parasites as we know very little indeed about their food requirements and may also know very little about where they live.

A different approach has therefore been adopted here. The overall aim is to explain the distribution and abundance of Acanthocephala in both space and time by searching for repeated patterns. It is essential, however, to appreciate that parasites form a nested hierarchy and, as with free-living animals, they can be studied at the individual, population and community levels. They can also be studied at different spatial scales, from local through regional to global, and over different temporal scales, from seasonal through annual to long term. It is therefore possible to pose a series of questions at each level and scale, and the answer to each question will form the subject of a separate chapter of this book. It is necessary to emphasise, however, that this book does not attempt to be comprehensive, but rather interpretative. Considerations of physiology and phylogeny, for example, are largely ignored in favour of a focus and emphasis on ecological issues and examples are selected to make particular points. This inevitably means that many other excellent examples have had to be omitted, but considerations of space require choices to be made.

The first stage in any ecological appraisal of a group of organisms is to learn as much as possible about their life cycles, and this is particularly important in respect of parasites. In the case of the Acanthocephala it is essential to question whether the number of hosts in the life cycle is really fixed at two, or whether any flexibility and variation in the number of hosts, by addition or deletion, is possible to assist transmission (Chapter 2). It is logical then to place the Acanthocephala in a biogeographical context and to question whether the species are equally distributed between all habitats and host groups and whether patterns in global and regional distribution can be detected (Chapter 3). A major influence on the ecology of all parasites is their host specificity and so it is essential to question how specific to each of their hosts are acanthocephalans and how much variation in specificity exists between species (Chapter 4). It then becomes apparent that there may be much more diversity at the molecular level than is apparent at the morphological one and this will raise yet more questions. A further question at the individual level is whether acanthocephalans are ever pathogenic to their hosts, or alter their host behaviour in any way, and if so under what circumstances and with what consequences and benefit (Chapter 5).

Moving to the population level, the key question is whether acanthocephalan populations are stable and regulated over time and space, or whether they are unstable and liable to local epizootics and, if so, under what circumstances (Chapter 6). Acanthocephalan species almost invariably co-occur with other species of parasite in host individuals and/or populations and so form a part of a community. At these levels it is pertinent to question whether species interact with other species of acanthocephalan or of other parasitic groups and compete for occupancy of niches and if so whether this inter-specific competition has wider ecological significance (Chapter 7). Acanthocephalans do not occur in isolation but are components of ecosystems. It is therefore important to query how they transfer from one ecosystem to another and/or colonise new localities (Chapter 8). As members of an ecosystem, they are also responsive to changes in that ecosystem and it is valid to question the extent to which they can act as indicators of such changes: it is equally important to appreciate the impact that they may have on an ecosystem and the extent to which they may affect it (Chapter 9). The answers to many of these questions may be unexpected and surprising and taken together not only aid our understanding of the ecology of the acanthocephalans but also go a long way to explaining their success as a group (Chapter 10). They also take us to the heart of understanding parasitism as a way of life as the Acanthocephala do seem to present a distinctive and successful pathway of host–parasite co-evolution *sensu* Anderson & May (1982) and they are indeed in the words of MacArthur (1972) 'worth studying'.

All these questions are basically ecological ones and all are answerable to a greater or lesser extent. Many of the answers will come from studies on aquatic species. This is not a bias on my part or a reflection of my particular interests: rather, it reflects the facts that the majority of acanthocephalan species have an aquatic life cycle and that since it is easier to obtain good samples from aquatic populations of fish and invertebrates, most ecological studies have focused on aquatic, especially freshwater, species. It is far easier to collect large and representative samples of fish and aquatic amphipods, for example, than of terrestrial cockroaches and avian raptors (or rhinoceroses!). Schmidt (1985) tabulated summary information on the life cycles of 125 acanthocephalan species, but listed intermediate hosts for only 80. Only 14 of these were terrestrial species. Any apparent bias towards particular species similarly reflects the ecological constraints and information available. Some species have been studied in many localities over wide geographical areas, for example *Pomphorhynchus laevis*, *Echinorhynchus salmonis* and *Macracanthorhynchus hirudinaceus*, and others such as *Mediorhynchus centrurorum* have been studied intensively in particular localities. A great deal of information is available on the biology of *Moniliformis moniliformis*, for example, but this reflects the ease with which it can be maintained in the laboratory, and there is still a dearth of information on its ecology. For the great majority of species, however, virtually nothing is known about their ecology.

Throughout this account the name of an acanthocephalan species as used by the authors in their publications is also the one used here, regardless of any subsequent changes or synonomising or transferring to a new genus. For example, I have used *Moniliformis moniliformis* and *M. dubius* and *Metechinorhynchus salmonis* and *Echinorhynchus salmonis* as used by the authors of the papers regardless of the current synonomy or genus to which the species is assigned. All definitions of ecological terms as applied in parasitology are to be found in Bush *et al.* (1997).

2

Life cycles and transmission

2.1 GENERAL STRATEGY

The Acanthocephala is a small, discrete and very distinct phylum. They share some similarities with the platyhelminths— for example with the cestodes in the absence of any alimentary tract and with the nematodes in having a body cavity, separate sexes and an invariable number of larval stages. Their particular combination of characteristics is, however, unique. They appear to be a very old and monophyletic phylum, and a fossil of what may be an acanthocephalan has been reported (Conway-Morris & Crompton, 1982). Nevertheless, their origins, relationships to other phyla and evolutionary history are still unclear, although current opinion suggests that they may be more closely related to the rotifers than to any other animal phylum (Garey *et al.*, 1996).

All adult acanthocephalans are obligatory endoparasites and live in the alimentary canal of a vertebrate. Each species exhibits a preference for a particular region of the alimentary tract (Crompton, 1973; Holmes, 1973) although its distribution may be more extensive than this and the preference may change with host and over time. The morphology of the adults is very similar indeed between species, especially in the internal organs. The proboscis bears hooks, the shape, number and arrangement of which differ between species and genera but the adaptive significance of the arrangement and of the specific differences is not understood. There is no clear relationship between hook arrangement and definitive host identity or even between preferred site in the alimentary canal. *Pomphorhynchus laevis* and *Acanthocephalus anguillae*, for example, can both use chub *Leuciscus cephalus* as a suitable definitive host and their distribution within the alimentary tract overlaps, yet their hook patterns are totally distinct (Fig 1.1); and all three species of *Acanthocephalus* can use eels *Anguilla anguilla* as a preferred definitive host. In addition to varying between species, hook arrangement and numbers can also vary within some species on a local (Brown, 1987) or regional (Buckner & Nickol, 1979) scale.

The acanthocephalans lack an alimentary tract and all uptake of nutrients takes place across the tegument. Feeding and nutrition have been reviewed by Starling (1985), from which it is clear that there are massive gaps in our understanding of their nutritional requirements. This is the situation in respect of most endoparasites and is in marked contrast to free-living organisms where a knowledge of the diet and food preference of an animal species is an enormous help in understanding its ecology and behaviour. We simply do not know exactly what any species of acanthocephalan requires from its host and so we do not know why some host species are suitable for a particular acanthocephalan species to grow and reproduce and others are not.

The most prominent interior organs are the reproductive ones (Fig. 1.2). The sexes are separate and females frequently attain a larger size than males. Reproduction is exclusively sexual: asexual reproduction and parthenogenesis are unknown in the phylum (Kennedy, 1993a). Site preference in the gut assists probability of contact and so mating between the sexes, but polygamy appears to be normal and males may move down the alimentary canal, fertilising females as they do so before being selectively lost (Crompton & Walters, 1972; Crompton, 1985). The ovaries break up into ovarian balls in which the eggs develop to maturity and the uterine bell in the females operates as an egg sorting device (Whitfield, 1970) which allows mature shelled acanthors to be released and immature ones to be retained. The whole process of reproduction has been reviewed in detail by Parshad & Crompton (1981) and Crompton (1985). Eggs are then released into the alimentary tract and exit the host in faeces. Rarely, an entire gravid female may be released with the faeces and the eggs are then liberated by the decay of the adult body, but this does not appear to be the normal method of egg release.

The shelled acanthors, normally and hereafter referred to as eggs, can be eaten by any invertebrate species, but if they are eaten by an arthropod that can serve as a suitable intermediate host the acanthella larva emerges in the intestine of the arthropod and burrows through its wall into the haemocoel. In an unsuitable host the eggs may be unable to hatch and so pass out in the faeces, or the acanthella may be unable to penetrate the intestinal wall or to develop further in the haemocoel, in which case it dies as a consequence of a host reaction. In a suitable host, it develops to the cystacanth stage which serves as a resting stage and is infective to the vertebrate definitive host. There is no variation in this life cycle other than in the identity of the intermediate host. All species without exception utilise an arthropod intermediate host: this may be a terrestrial beetle or cockroach if the definitive host is a terrestrial bird or mammal, or it may be a crustacean (Fig. 1.3) or ostracod for a freshwater species or a crab for a marine one, but all species go through the same larval stages and require only a single intermediate host. This uniformity of life cycle might suggest that there is little flexibility in development or host use, but in reality there can be a high degree of variability in numbers and identity of hosts used.

2.2 THE ADULT STAGE

The lifespan of adults within their definitive hosts is very variable (Table 2.1). Some species such as *Macracanthorhynchus hirudinaceus* have a much longer patent than pre-patent period and so can live for well over a year in pigs, whereas other species such as *Polymorphus minutus* have pre-patent and patent periods of similar lengths and live for only between one and two months in ducks. A few species, including *Fessisentis nectuorum*, live for only very short periods (days) in their definitive amphibian host (Nickol & Heard, 1973). Longevity and the relative lengths of the pre-patent and patent periods do not always appear to bear a close relationship to the identity of the definitive host. Clearly survival for long periods is only possible in a host such as a pig that itself has a long lifespan, and the relatively constant conditions of the intestine in a homeotherm host may permit a continuous and extended period of reproduction. By contrast the brief survival of *F. nectuorum* may well relate to the fact that it parasitises only larval salamanders and cannot survive the metamorphosis of its host. This species also has a very short pre-patent period as there is some precocious sexual development in the cystacanth stage. Other species infecting amphibians which can survive the metamorphosis of their hosts have longer lifespans (Rankin, 1937; Avery, 1971).

It is not immediately clear, however, why acanthocephalans of birds should exhibit shorter lifespans and why *Polymorphus minutus* should survive for such a short period in a domestic duck while *Profilicollis botulus* survives for at least twice as long in an eider duck. It is

Table 2.1. *Life history data on some species of acanthocephalans*

Species	Duration (months)		Total eggs per female	References
	Cystacanths	Adult		
Archiacanthocephala				
Macracanthorhynchus hirudinaceus[a]	2 minimum	14	24×10^6	Kates (1942, 1944)
Moniliformis dubius[a]	2 minimum	4.6	60×10^4	Arnold & Crompton (1987)
Mediorhynchus centurorum[a]	6	4	ND	Nickol (1977)
Palaeacanthocephala				
Polymorphus minutus[b]	15	1.7	13×10^3	Crompton & Whitfield (1968)
Profilicollis botulus[b]	36	3.3	ND	Thompson (1985a,b)
Acanthocephalus lucii[c]	8	6.1	60×10^4	Komarova (1950) Serov (1984)
Echinorhynchus truttae[c]	7	2.9	ND	Awachie (1966)

[a]Terrestrial intermediate and definitive hosts.
[b]Aquatic intermediate host and bird definitive hosts.
[c]Aquatic intermediate host and fish definitive host.
ND, no data.
Source: Modified from Kennedy (1993a).

probable that *P. minutus* is basically a parasite of migratory ducks, and its *Gammarus* intermediate host generally has a lifespan of a year or less. Under these circumstances it would be advantageous to have a short patent period and so a rapid production of eggs in a bird that spends only a relatively short time in any one locality. Eiders, by contrast, are not truly migratory but are displaced seasonally within their breeding range, and the intermediate host of *P. botulus* is a crab with a lifespan of well over a year. It is less clear, however, why *Acanthocephalus lucii* should require a much longer patent period in perch than *Echinorhynchus truttae* does in trout or why the former species should survive twice as long in its fish host as the latter, since both fish species are capable of surviving for several years and both use crustacean intermediate hosts with lifespans of less than a year on average. Parasites in fish hosts can generally survive seasonal periods of host starvation, whereas parasites of homeothermic hosts cannot.

There is a real paucity of data on longevity, fecundity and patency periods, due largely to the fact that eggs are very difficult – to the extent of being often impossible – to detect in host faeces qualitatively or quantitatively. It can be tentatively suggested that few species survive in their definitive host for longer than a year and those probably only in some homeotherms. Even then, individual variation in egg production is high, as in *M. dubius*, and there are large inequalities in individual fecundity (Crompton *et al.*, 1972; Dobson, 1986). There are also differences in fecundity between parasites in different individual hosts and these may in turn relate to individual differences between hosts in their diet (Crompton, 1985; Crompton *et al.*, 1988). Intraspecific competition will probably also lead to a reduction in parasite fecundity, but this has yet to be demonstrated for the reasons discussed above. These differences are in large measure phenotypic, and are superimposed on genotypic differences which are adaptive and permit a degree of flexibility within the apparent rigidity of the life cycle. In terms of life cycle strategy, most parasites can be considered to lie closer to the r end of the $r - K$ continuum as defined by Pianka (1974) and Esch *et al.*, (1977), but the Acanthocephala are very difficult to locate along this continuum as they exhibit features of both r and K selection.

2.3 EGGS

The uterine bell mechanism ensures that eggs are only released from the adult when they are fully mature and infective to the intermediate host (Whitfield, 1970). Most data on egg survivorship has come from laboratory investigations, in view of the virtual impossibility of detecting eggs in field conditions. Such data as exist indicate considerable variation in the survival time of eggs even in the laboratory (Table 2.2). These data might at first suggest that eggs survive for a shorter period in terrestrial conditions, but more recent and extensive studies have shown that eggs of *Moniliformis dubius* can survive for much longer periods, up to two years (Arnold & Crompton, 1987) and eggs of *Macracanthorhynchus hirudinaceus* can survive and remain infective for up to three years (Kates, 1942, 1944). Instead, it can be tentatively suggested that eggs of terrestrial species survive for longer periods than those of aquatic species. In all probability survivorship of eggs of

Table 2.2. *Survivorship of acanthocephalan eggs under experimental conditions*

Species	Temperature (°C)	Period (months)	Reference
Archiacanthocephala			
Macracanthocephalus hirudinaceus[a]	5	18	Kates (1942)
Moniliformis dubius[a]	5	1	Edmonds (1966)
Palaeacanthocephala			
Polymorphus minutus	15	6	Petrochenko (1956)
Pomphorhynchus bulbocolli	4	6	Jensen (1952)
Prosthorhynchus formosus	5	10	Schmidt & Olsen(1964)
Leptorhynchoides thecatus	4	9	De Giusti (1949)
Eoacanthocephala			
Octospinifer macilentis	4	9	Harms (1965)
Neoechinorhynchus rutili	4	6	Merritt & Pratt (1964)

[a]Infects terrestrial intermediate hosts.
Source: From Crompton (1970).

all species is time dependent and decline in infectivity is exponential, as with other parasites of other phyla, but there are no data on this for acanthocephalans.

Eggs are certainly resting and resistant stages. Eggs of terrestrial species may be exposed to environmental conditions that fluctuate more extensively than those in fresh water, and it is not too surprising to find that eggs of *M. hirudinaceus* can withstand freezing down to −16 °C in water or drying for up to 140 days and warming up to 26 °C for 265 days (Kates, 1942). Even some eggs of the aquatic species *Polymorphus minutus* remain infective to their intermediate host after being frozen for 2.5 months at −22 °C (Hynes & Nicholas, 1963). The ability of eggs to survive harsh conditions is clearly adaptive and enables parasites to survive in a locality over winter when their hosts are absent or inactive.

Eggs are transmitted to the intermediate hosts by ingestion, but they do have some adaptations to improve the probability of transmission success. In some aquatic species, the outermost egg membrane is lost and a fibrillar coat releases filaments. This may anchor the eggs to each other in clusters of five to ten and entangle them in aquatic vegetation, as is known to occur in *Acanthocephalus jacksoni* according to Oetinger & Nickol (1974). In *Leptorhynchoides thecatus* the fibrils tangle with algae, which form the diet of their intermediate amphipod hosts. That transmission is facilitated is evidenced by the fact that amphipods feeding on algae and eggs exhibit higher levels of infection than amphipods feeding on eggs alone (Uznanski & Nickol, 1970). By contrast, the membrane of eggs of *Pomphorhynchus bulbocolli* does not break down, fibrils are not released and the eggs fall to the bottom of the water body. Both these latter species use the same intermediate host, *Hyalella azteca*, and it has been demonstrated experimentally that infection levels of *L. thecatus* are significantly higher than those of *P. bulbocolli* (Barger & Nickol, 1998,1999). Another adaptation has been demonstrated by Wongkham & Whitfield (2004), who have similarly shown that the uptake of water by eggs of *Pallisentis rexus* causes the eggs to expand, which reduces their density and enables them to float, thus bringing them into the orbit of their plankton-feeding intermediate hosts.

2.4 INTERMEDIATE HOSTS AND CYSTACANTHS

All acanthocephalans require an arthropod intermediate host, in the haemocoel of which the acanthella develops into the cystacanth stage. The cystacanth is the stage that is infective to the definitive host, and transmission to the next host can only be accomplished by ingestion of an infected intermediate host. However, the cystacanth may have to develop to some degree before it is fully infective. Cystacanths of *Leptorhynchoides thecatus* younger than 26 days old, for example, are unable to infect rock bass: between 26 and 29 days old they can infect rock bass but are unable to remain in the intestine and pass through the intestinal wall to locate in an extra-intestinal site; and only when they are older than 30 days are they able to remain in the intestine and develop further (De Giusti, 1949; Nickol, 1985). It seems probable that once developed a cystacanth survives as long as the intermediate host that harbours it. Because many species of arthropods are short lived, i.e. less than a year old, it seems probable that few cystacanths survive for longer than a year, although cystacanths of *Profilicollis botulus* which infect shore crabs (which live for one or more years) may survive as long as their hosts (Table 2.1). Cystacanths can be considered to be resting stages, again facilitating survival of the parasite over seasonally unfavourable climatic conditions. In rare cases they may undergo a true diapause: development of pre-cystacanths of *Polymorphus marilis* in *Gammarus lacustris* are halted at cold winter temperatures.

Development is then resumed at warmer temperatures, but the rate of resumption is directly related to the length of time spent at the low temperatures, thus ensuring that there is no premature development of the larvae in warm spells over winter (Tokeson & Holmes, 1982).

The identity of the intermediate host relates both to habitat and to the class of Acanthocephalan. Archiacanthocephalans are terrestrial and as adults infect terrestrial vertebrates: the intermediate hosts of the species in this class are terrestrial arthropods belonging to several insect orders including Coleoptera, Dermaptera and Orthoptera, and some species may use Myriapoda. Members of the aquatic Eoacanthocephala generally use aquatic crustaceans, chiefly Ostracoda, as their intermediate hosts, although a few species have been reported from Copepoda and Amphipoda. The Palaeacanthocephala contain both aquatic and terrestrial members: the former generally use aquatic isopods or amphipods as intermediate hosts and the latter may do the same or use terrestrial amphipods. Many species, whether terrestrial or aquatic, employ as intermediate hosts species that are key organisms in a food web.

In any locality, an intermediate host may harbour more than one species of acanthocephalan. In a single lake in Canada, for example, *Gammarus lacustris* was found by Denny (1969) to be used by three species of *Polymorphus*, all of which may use the same definitive host species. Elsewhere, *Hyalella azteca* has been reported to be co-infected with *Pomphorhynchus bulbocolli* and *Leptorhynchoides thecatus*, although these two species used different definitive hosts (Barger & Nickol, 1999). Similarly, Nickol *et al.* (2002a) reported the use of the fiddler crab *Uca rapax* by *Hexaglandula corynosoma* and *Arhythmorhynchus frassoni*, but again each species used different definitive hosts, a night heron in the first case and a water rail in the second. However, co-infections in single individual hosts may be fairly uncommon, as prevalence levels in intermediate host populations are often very low: Denny (1969) for example reported prevalence levels of 12.7%, 0.8% and 0.8% for the three species of *Polymorphus* in *Gammarus lacustris* in Cooking Lake, Amin (1978) reported infection levels of 0.1% and 0.01% for two species of acanthocephalan in *Pontoporeia affinis* in Lake Michigan and Ashley & Nickol (1989) found <0.7% of *Hyalella azteca* infected in one locality.

This use of an intermediate host by more than one species prompted Bush *et al.* (1993) to suggest that intermediate hosts might actually provide a source community to the definitive host and not just a source population of a single species. They pointed out as an example that a single definitive host feeding on an intertidal crab could be colonised simultaneously by several species of parasite. No really convincing examples of this have been reported for acanthocephalans although it would seem possible that a single duck could acquire more than one species of *Polymorphus* in this way, and eels in Lough Derg could certainly acquire both *Acanthocephalus anguillae* and *A. lucii* simultaneously from *Asellus aquaticus* (Kennedy & Moriarty, 1987). It has also been suggested that parasites unable to modify the behaviour of their intermediate host to facilitate transmission to their definitive host would be advantaged if they were able to preferentially infect hosts already infected by a debilitating parasite (Thomas *et al.*, 1997). However, no example of this has been reported from acanthocephalans, perhaps because most species are able to modify the behaviour of their intermediate hosts themselves.

2.5 PARATENIC HOSTS

Specialisation of a species for an intermediate host is both an advantage and a disadvantage, in that it effectively canalises the cystacanth into a narrow direction. It can only be acquired by an organism that eats the infected intermediate host. The nature of a food web is that in reality many organisms may eat infected intermediate hosts and some of them may not be suitable hosts for the parasite. In many species the parasite may then die, but in others the parasite may be able to survive even if it does not grow or develop further, and if this host is in turn is eaten by a definitive host, then the parasite may continue development. This type of host is generally refered to as a paratenic (or transport) host. It is never an obligatory host in the life cycle but only a facultative one, and it may or may not facilitate transmission to the definitive host. It can be interpolated into the life cycle at any stage, but in the case of acanthocephalans it is always a vertebrate and is to be found between the intermediate and definitive hosts (Fig. 2.1).

Paratenic hosts are to be found in many acanthocephalan life cycles. On reaching such a host, the cystacanth may partially evert and move into the body cavity of the vertebrate, where it is often to be found attached to the mesenteries and/or partially encysted. Unfortunately, it is often very difficult to determine whether a host with such parasites in the body cavity is actually a paratenic host, or an accidental, unsuitable potential definitive host, in whose body cavity acanthocephalans may also occur. Even in suitable definitive hosts it is not uncommon to find mature adult acanthocephalans in the

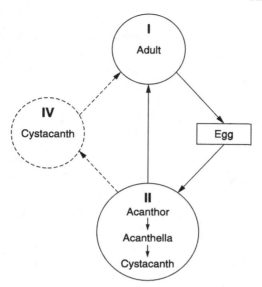

Fig. 2.1 Development of the life cycle of an acanthocephalan. Host types are indicated by roman numerals (I, definitive host, II, intermediate host, IV, paratenic host). Solid lines represent the basic life cycle of obligatory hosts, and dotted lines indicate the incorporation of an extra facultative host in the life cycle. Stages within hosts are enclosed in circles and the free-living stage is enclosed in a rectangle. (From Bush *et al.*, 2001).

intestine and partially excysted ones in the body cavity and its organs. Acanthocephalans look the same in both types of host, and this has caused considerable confusion between accidental and paratenic hosts in the literature as the only way of identifying a paratenic host as such is to determine whether the parasite will resume development if transferred to a suitable definitive host. However, field data may provide good indications. Thus, the finding by Ewald *et al.* (1991) of *Centrorhynchus* larvae in *Sorex* species on a regular basis and at high intensity levels suggests that this is not merely an accidental occurrence but that the shrews are paratenic hosts, probably of *C. aluconis* which mature in owls.

The extent to which paratenic hosts are incorporated into acanthocephalan life cycles varies with the class of acanthocephalans and the species. The table of Schmidt (1985) provides an excellent summary of the use of paratenic hosts by species of Acanthocephala, and the data in Table 2.3 are abstracted from this. In general, few species of Palaeacanthocephala use paratenic hosts, *L. thecatus* and *Corynosoma* species being exceptional in this respect. By contrast,

Table 2.3. *Use of hosts by selected species of Acanthocephala*

Species	Intermediate		Paratenic		Definitive		
	N.sp.	N.ge.	N.sp.	N.ge.	N.sp.	N.ge.	
Archiacanthocephala							
Macracanthorhynchus hirudinaceus	33	26	0	0	3	2	M
M. catulinus	6	6	18	18	12	11	M
M. ingens	7	5	9	8	5	5	M
Mediorhynchus centurorum	1	1	0	0	4	4	B
M. grandis	5	5	1	1	5	5	B
M. papillosus	5	5	0	0	13	13	B
Moniliformis moniliformis	8	6	0	0	3	1	M
M. clarki	2	1	0	0	13	11	M
Palaeacanthocephala							
Acanthocephalus anguillae	1	1	0	0	29+	27+	F
A. clavula	9	7	0	0	25+	22+	F
A. dirus	6	4	0	0	21+	21	F
A. lucii	1	1	0	0	15+	15	F
Leptoryhnchoides thecatus	2	1	4	3	41+	41+	F
Pomphorhynchus bulbocolli	2	2	0	0	38+	38	F
P. laevis	1	1	0	0	16+	14	F
Echinorhynchus gadi	5	5	0	0	33	24	F
E. salmonis	1	1	0	0	37	23	F
Polymorphus minutus	11	3	0	0	15+	15+	B
Corynosoma semerme	u	u	26	18	13	7	M
Eoacanthocephala							
Neoechinorhynchus rutili	1	0	1	1	36+	36+	F
N. cylindratus	1	1	3	2	27+	27+	F
N. emydis	2	2	2	1	12	6	R
Octospiniferoides chandleri	2	2	0	0	2	1	F
Paratenuisentis ambiguus	1	1	0	0	1	1	F

N.sp., number of species; N.ge., number of genera; M, mammal; B, bird; R, reptile; F, fish; n+, minimum number; u, not determined in study.

Source: Based on, and modified from, Schmidt (1985).

many, but by no means all, species of Archiacanthocephala and Eoacanthocephala do use paratenic hosts. Usage, however, can differ markedly between species in the same genus: for example, *Macracanthorhynchus hirudinaceus* does not use paratenic hosts but *M. catulinus* and *M. ingens* do (Table 2.3).

Paratenic hosts also provide a very important means of overcoming the rigidity of a two-host life cycle and the limitations of a single intermediate host life cycle by enabling acanthocephalans to bridge or jump trophic levels. For example, cystacanths of *Corynosoma* species utilise marine amphipods as intermediate hosts, but the adults of species such as *C. semerme* are found in seals. Amphipods are not a primary dietary item of seals, which are basically piscivorous. By contrast, many species of fish do feed on amphipods. It is therefore not surprising to find that many species of fish serve as paratenic hosts to *C. semerme* (Helle & Valtonen, 1981; Valtonen & Crompton, 1990). Similarly, some species of *Centrorhynchus* utilise aquatic snakes and amphibians as paratenic hosts to jump habitats and bridge the gap between aquatic invertebrate intermediate hosts and aquatic predatory birds such as herons and more terrestrial raptors. Even in fresh waters, use of a paratenic fish host bridges the gap between, for example, an ostracod intermediate host and a predatory piscivorous fish host.

The adults of many species of *Bulbosoma* infect whales and the intermediate host is a species of marine plankton. Costa *et al.* (2002) have suggested that two cycles may exist in the waters off Portugal: a pelagic cycle with a pelagic crustacean as intermediate host, and mackerel and dolphins as paratenic hosts, and a deep water cycle with a crustacean as intermediate host and a benthic fish species as paratenic host. Paratenic hosts may also bridge the habitat gap between water and land, or land and air. In a similar manner, *Macracanthorhynchus catulinus* uses a variety of snakes, lizards and frogs to infect its canine and mustelid definitive hosts. Lizards are less frequently infected with acanthocephalans than other classes of vertebrates (see Chapter 3), but they can be commonly employed as paratenic hosts for parasites that mature in birds and mammals (Sharpilo *et al.*, 2001). A particularly clear indication of the importance of paratenic hosts in the epizootiology of acanthocephalans has been provided by Elkins & Nickol (1983) for *Macracanthorhynchus ingens* in Louisiana. The first intermediate host is a species of woodroach, and cystacanths from woodroaches are able to develop viscerally in colubrid and viperid snakes, both aquatic and terrestrial, or to develop into the adult stage in racoons. The cystacanths from snakes are capable of maturing in the gut of racoons. Both pathways occur in nature. Cystacanths can be found in armadillos, but they then degenerate so this is an accidental host and not a true definitive one. Racoons are the preferred host of this species, and can and do eat both insects and snakes. In the area studied, the snakes tended to live in damp areas under logs, a habitat in which roaches were common and in which racoons also hunted.

In effect, those species of acanthocephalans that use a paratenic host have escaped the rigidity and constraints of the two-host cycle. They have added an extra host, albeit a facultative one, at a different trophic level in the life cycle, which enables them to ascend trophic levels and so climb up food chains. Moreover, because paratenic hosts are not obligatory they can introduce a degree of local flexibility into the cycle by increasing range in host usage, thereby facilitiating transmission of acanthocephalans to their definitive host(s).

2.6 POST-CYCLIC TRANSMISSION

Post-cyclic transmission is yet another means by which the Acanthocephala can escape from the rigidity of the single intermediate host life cycle. For a long time it had been difficult to understand how predators could harbour heavy infections of an acanthocephalan species when paratenic hosts were not utilised and when the intermediate host formed only an insignificant part of their diet. Post-cyclic transmission, described by Nickol (1985) as 'When ingested as adults within their definitive hosts some acanthocephalans survive and parasitise the predator', can explain this problem. The parasites, instead of dying or being lost, simply reattach to the intestine of the predator, which may be the same as or a different species from the definitive host, and resume growth or reproduction. At that time little was known about the significance of post-cyclic transmission in acanthocephalans, although Nickol listed five species from which it had been reported.

Eighteen years later, Nickol (2003) reported that it had now been demonstrated to occur in ten species of acanthocephalans and from species representative of each class of Acanthocephala, and he believed that its significance was still underestimated. In the following year, McCormick & Nickol (2004) reported its occurrence in an eleventh species. The ease with which it could occur in those species in which it had been experimentally demonstrated suggested that it could be of widespread importance in nature. To be of importance, of course, the transferred adults must survive long enough to become sexually mature, or if mature must continue to reproduce after transfer. This seems often to be the case (Table 2.4), although in species such as *Pomphorhynchus laevis* in which attachment to the definitive host involves a fibrous tissue encapsulation of the proboscis bulb, it is only less mature specimens in which the bulb is not fully developed that are able to transfer in this way (Kennedy, 1999). The most recent example reported by McCormick and Nickol (2004) demonstrates its value in understanding field data. Infection levels of *Paulisentis missouriensis* increased with the size of creek chub, but this was not due to increased consumption of the intermediate host, which is

Table 2.4. *Post-cyclic transmission of acanthocephalans in the laboratory*

Species transferred	Stage	Time (days)	Growth	Reference
Archiacanthocephala				
Macracanthorhynchus catulinus	A	u	Yes	Farzaliev & Petrochenko (1980)
Moniliformis moniliformis	A	21	Yes	Moore (1946)
Palaeacanthocephala				
Acanthocephalus ranae	u	u	u	Bozkov (1980)
Acanthocephalus tumescens	u	28	Yes[a]	Rauque *et al.* (2002)
Acanthocephaloides propinquus	u	48	u	Buron & Maillard (1987)
Echinorhynchus salmonis	u	84	u	Hnath (1969)
Pomphorhynchus laevis	A[b]	31	Yes	Kennedy (1999)
Eoacanthocephala				
Neoechinorhynchus cristatus	A	7	u	Uglem & Beck (1972)
Neoechinorhynchus rutili	u	10	u	Lassiere & Crompton (1988)
Octospiniferoides chandleri	u	7	u	DeMont & Corkum (1982)
Paulisentis missouriensis	A	14	Yes[c]	McCormick & Nickol (2004)

[a]Some females gravid after transfer, but status before undetermined.
[b]Young adults transferred, but older ones with fibrotic capsule around proboscis did not.
[c]Transferred adults attained sexual maturity.
A, adult; u, not determined in study. Duration times are the minimum.
Source: From Nickol, (2003) and McCormick & Nickol (2004).

a copepod, or to the use of a paratenic host as none is known in the cycle. In fact the explanation lay in cannibalism, as larger fish consumed more fish of their own species.

The potential significance of this form of transmission is that, as with the use of a paratenic host, the parasite can move up trophic levels and be transferred to new host groups at these levels that might otherwise be unavailable as hosts. This form of transmission can in addition accentuate aggregation of parasites in hosts and explain some occurrences in odd hosts which do not eat the intermediate or paratenic hosts. It seems likely that many more examples will be reported and that it may prove to be widespread in nature, even though hard to detect with certainty outside the laboratory. Thus usage of both paratenic hosts and post-cyclic transmission enables acanthocephalans to escape the constraints of a two-host cycle and acutally employ a four-host one.

2.7 VARIATION IN LIFE CYCLES

As the previous examples have demonstrated, the invariability of the acanthocephalan life cycle is more apparent than real. Lengths of pre-patent and patent periods are adapted to the climatic conditions in which the parasite lives and to host movements in and out of a locality and other seasonal activities. Eggs and cystacanths are resting stages, in which the parasite can survive unfavourable conditions and/or host's seasonal absences. The use of paratenic hosts in the life cycle and the ability to transmit post-cyclically effectively increases the number of hosts in the cycle and so allows escape from the apparent rigidity of a single intermediate host. This enables the acanthocephalans to jump levels in the food chain, i.e. to climb trophic levels, and also to transfer from one medium such as water to another such as land or air. In reality, the life cycles are surprisingly flexible, especially when it is considered that some species can use a wide variety of species for both intermediate and paratenic hosts.

What this means is that in some cases a parasite species may appear to have quite different life cycles in different habitats. The suggestion of Costa *et al.* (2000) that *Bolbosoma vasculosum* may have both planktonic and benthic life cycles in different sea areas has already been discussed. This, however, is not the only example. Walkey (1967) investigated the ecology of *Neoechinorhynchus rutili* in a pond in southern England, where he believed that the life cycle involved a single intermediate host, a species of ostracod, and that the three-spined stickleback *Gasterosteus aculeatus* was the definitive host. As there had been records of larval parasites from the larvae of the megalopteran *Sialis lutaria*, he attempted to infect them experimentally, but failed. In Scotland, Lassiere (1988) found adult *N. rutili* to be common in brown trout *Salmo trutta* as well as sticklebacks, and also in larvae of *S. lutaria*. In her localities trout became infected by eating the infected *Sialis*, which appeared to be functioning as a paratenic host. In addition, Lassiere & Crompton (1988) were able to demonstrate experimentally post-cyclic transmission between infected sticklebacks and rainbow trout. It would seem to be only a small step for these phenotypic, ecological host preferences to be incorporated into the gene pool and so for new strains to evolve.

This last example is not the only strange feature of *Neoechino-rhynchus* species life cycles. Adult *N. emydis* use freshwater turtles as definitive hosts and ostracods as intermediate hosts. This species is also reported to use a species of snail as a paratenic host, a choice unique among acanthocephalans. Even more surprising is the fact that Hopp (1954) was able to show that cystacanths from snails were infective to turtles, but not cystacanths from ostracods. This would indicate that the snail is not, in fact, a paratenic host at all but a true intermediate host which is required for the cystacanth to complete development. This would therefore seem to be the only example known to date in which an acanthocephalan life cycle requires two inter-mediate hosts, one of which is not an arthropod. It also provides a further example of the diversity of acanthocephalan life cycles.

3

Biogeography and distribution

3.1 PHYLOGENETIC INFLUENCES

There is widespread acceptance that the phylum Acanthocephala can be divided into three major classes (Amin, 1985a), and some authorities also recognise a fourth, very small, class. The characterisation of these taxa is based on a number of structural features such as body size, number of cement glands and trunk spination and also on a number of biological features including habitat and identity of intermediate and definitive hosts.

The most important of these biological features in relation to the three main classes are summarised in Table 3.1. The Archiacanthocephala are truly terrestrial: they use terrestrial insects and myriapods as intermediate hosts and predatory birds and mammals as definitive hosts, and many, though not all, species also have the ability to use paratenic hosts (Schmidt, 1985), particularly reptiles and amphibians. By contrast the Palaeacanthocephala are mostly aquatic and use aquatic arthropods as intermediate hosts. They show the greatest diversity of all three classes in their use of definitive hosts. Many species use teleost fish, but others use amphibians, birds and mammals. These birds and mammals generally show strong aquatic links and species such as ducks, seals and whales serve as hosts. The use of paratenic hosts is not very common, but when they are employed it is often to bridge levels in food chains, for example in the use of fish to bridge the levels of crustacean intermediate hosts and piscivorous seals as definitive hosts. A few species, for example *Plagiorhynchus cylindraceus*, are completely terrestrial but this species utilises a terrestrial member of a primarily aquatic group as its intermediate host, namely terrestrial isopods of the genus *Armadillidium*. The Eoacanthocephala are all completely aquatic: they

Table 3.1. *Summary of the ecological characteristics of the main[a] classes of Acanthocephala*

	Class		
	Archiacanthocephala	Palaeacanthocephala	Eoacanthocephala
Habitat	Terrestrial	Most aquatic	Aquatic
Intermediate hosts	Insects, myriapods	Aquatic arthropods	Ostracods
Paratenic hosts	Common	Not very common	Uncommon
Definitive hosts	Mainly predatory birds and mammals	Fish, amphibians birds and mammals	Fish, amphibians and chelonians

[a]The fourth class, Polyacanthocephala, is very small. Species are aquatic and adults infect South American caimans. Their life histories are not known.

use ostracods and copepods as intermediate hosts and fish, amphibia and reptiles, particularly turtles, as definitive hosts. Some species also use paratenic hosts to bridge levels in food chains, for example between ostracods and piscivorous turtles. The Polyacanthocephala are a small, isolated, aquatic group that infect South American caimans as definitive hosts.

Overall, there are approximately 1000 species in the phylum, of which around 57% are in the Palaeacanthocephala and the remaining species are divided equally between the other two major classes. The species are grouped into 125 genera and each class contains a few genera that are species rich and a few that are species poor (Table 3.2). The pattern of distribution of species in genera is very similar between classes and the occurrence of both species-rich and species-poor genera appears to be independent of host and habitat. The richest genera of Archiacanthocephala, *Mediorhynchus* and *Oligacanthorhynchus*, parasitise terrestrial birds; the richest genera of Palaeacanthocephala, *Centrorhynchus* and *Acanthocephalus*, parasitise terrestrial and aquatic birds and mammals, and fish and amphibia respectively, and the richest genus of Eoacanthocephala, *Neoechinorhynchus*, parasitises fish and turtles. At the other end of the scale, 37% of the genera are monospecific and a further 17% contain only two species. The proportion of species-poor genera is lowest in the Archiacanthocephala but higher and similar in the other two classes.

Table 3.2. *Frequency distribution of species per genus of Acanthocephala*

Class	No. of		Number of species per genus										
	Ge.	Sp.	1	2	3	4	5	6–10	11–20	21–30	31–40	41+	Max.
Archiacanthocephala	13	168	1	0	3	0	0	5	1	1	1	1	45
Palaeacanthocephala	84	573	22	15	6	5	3	11	3	3	2	4	75
Eoacanthocephala	28	167	13	6	4	1	1	0	1	1	0	1	74
Total numbers	125	908	46	21	13	6	4	16	5	5	3	6	
%			37	17	10	5	3	13	4	4	2	5	

Ge., genera; sp., species.

Source: Data from Amin (1985a) with Polyacanthocephala and all species *incertae sedis* omitted.

Table 3.3. *Distribution of species of Acanthocephala by definitive host and habitat*

(a) By definitive host

	Host class				
	Fish	Amphibia	Reptiles	Birds	Mammals
Percentage of species	38.8	2.0	1.8	36.9	20.5

(b) By habitat

	Habitat		
	Freshwater	Marine	Terrestrial
Percentage of species	37.9	24.8	37.2

Source: Data from Yamaguti (1963). Polyacanthocephala omitted.

The distribution of species by definitive host utilisation (Table 3.3a) shows that fish and birds are the most widely used definitive hosts, followed by mammals. It is interesting that the oldest group of vertebrates, the fish, is not utilised by significantly more species than the youngest groups of birds and mammals. Although some species do use amphibians and reptiles as definitive hosts, the proportions doing so are very similar and very low. It would therefore seem that some factor other than age of host

class is influencing the utilisation of definitive hosts by acanthocephalan species. This factor appears to be habitat (Table 3.3b). While
a minority of 37.2% of species infect truly terrestrial definitive hosts,
the greater majority of 62.7% infect aquatic hosts and the majority of these utilise freshwater vertebrates. The acanthocephalans are
primarily an aquatic group of parasites even though they have
successfully invaded the land.

The one group of hosts that is not parasitised by acanthocephalans is the elasmobranchs. There is a record of an *Acanthocephaloides*
species from a *Raja* species, but this may well represent an accidental
infection rather than evidence of a ray being used as a normal
definitive host (Williams & Jones, 1994). Ancient teleosts such as species
of *Amia* and *Acipenser* are regularly utilised by species of acanthocephalans as definitive hosts (Aho *et al.*, 1991, and Ibragimov, 1985,
respectively). Over evolutionary time, it seems probable that the
herptiles have captured *Neoechinorhynchus* species from fish and aquatic
birds. Birds, and mammals (Connell & Corner, 1957), have similarly
captured *Polymorphus* species from fish, while *Corynosoma* species may
have been captured from birds by mammals (Bush *et al.*, 1990). As noted
for other groups by these same authors, fish have never captured
acanthocephalans from other groups as captures always seem to have
proceeded from the older group to the younger ones. Multiple
congeners may also occur among the acanthocephalans (Fig. 3.1),
chiefly in turtles (Aho *et al.*, 1992) and seals (Valtonen *et al.*, 2004), but
these are not truly species flocks as identified in other groups of
parasites by Kennedy & Bush (1992).

All these generalisations and values have to be treated with
some caution, as knowledge of the acanthocepalan biology is heavily
biased towards a very few species. For several species there exists only
a single record of their occurrence, and for others nothing at all
is known about their life cycle or the identity of their intermediate
host. Information is particularly poor in respect of some marine and
terrestrial species, because of difficulties of sampling and/or experimentally confirming life cycles. However, it does seem clear that
the Acanthocephala are primarily, and probably originally, an aquatic
group of parasites, and while the range of intermediate hosts
is confined to arthropods, both terrestrial and aquatic, the range of
vertebrates that can be utilised as definitive hosts is very diverse
indeed, and incorporates all the major groups other than the
elasmobranchs.

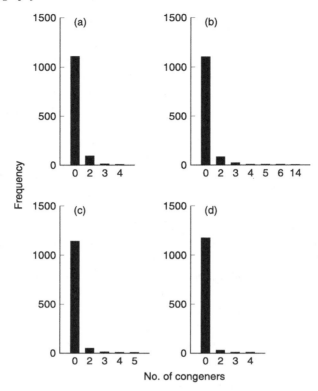

Fig. 3.1 Frequency distribution of the maximum number of congeneric species/study/parasite group (all host groups combined) for (a) Digenea, (b) Cestoda, (c) Nematoda, (d) Acanthocephala. (From Kennedy & Bush, 1992.)

3.2 BIOGEOGRAPHICAL INFLUENCES

3.2.1 Global patterns

The distribution of acanthocephalans as a group is worldwide. They have been reported from all continents and from all oceans including the Southern Ocean off the South Shetland Islands in Antarctica (Zdzitowiecki, 2001; Sures & Reimann, 2003) and the Arctic Ocean off Greenland (Arai, 1989), and freshwater species have been found as far north as Baffin Island (Curtis, 1979). Within the oceans, they have been recorded in surface-dwelling cetaceans (Balbuena & Raga, 1993, 1994) and from deep-sea, hydrothermal vents (Buron & Morand, 2004). However, within this broad picture it is possible to detect some patterns in global richness.

It is common with many groups of organisms to find a greater species diversity in the tropics. This is true of marine monogenean parasites (Rohde, 1993), but it is not the case with the acanthocephalans. A comparison of the richness of acanthocephalans in freshwater fish between northern temperate Nearctic (Canada) and several tropical localities by Choudhury & Dick (2000) revealed that acanthocephalans are in fact commoner in helminth communities in Canada than in the tropics (Table 3.4). They stressed the difficulty in making direct comparisons because of complications due to host and parasite lineages and host speciation especially in the tropics, but they tentatively further concluded that acanthocephalans were noticeably rare in parasite communities of fish in the neotropics, West Africa and Bangladesh, i.e. in tropical regions around the world, but were commoner in temperate regions. A re-examination of their data by Poulin (2001) controlling for sampling effort, host body size and phylogenetic contrasts nevertheless confirmed their conclusions that temperate helminth communities, including those of acanthocephalans, were still richer. Although Choudhury & Dick believed that the explanation for these results running counter to the prediction of greater richness in the tropics lay in differences in host diet in different regions, Poulin considered that the real reasons were still unknown.

Reports from other host groups show a similar pattern. The number of acanthocephalan species infecting pinnipeds and cetaceans at different latitudes is higher in the boreal and antiboreal regions than in the tropics and the Arctic and Antarctic regions (Table 3.5).

Table 3.4. *Richness of acanthocephalans in helminth communities of freshwater fish*

	Frequency distribution (%) of species in component communities						No. of cc	No. of ss
	0	1	2	3	4	5+		
Canada								
Acanthocephala	28.5	36.1	19.2	10.7	5.3	0	130	47
Other helminths	0.8	7.8	7.0	12.5	15.6	56.2		
Tropics								
Acanthocephala	71.8	24.3	3.1	0.7	0	0	160	118
Other helminths	16.3	20.0	21.2	12.5	13.7	15.7		

cc, component communities; ss, fish species.
Source: Data from Choudhury & Dick (2000).

Table 3.5. *Helminth species of pinnipeds and cetaceans at different latitudes*

Latitude	Acanthocephalans	Trematodes	Cestodes	Nematodes	Total
Arctic	3	6	7	9	25
Boreal	9	30	24	31	94
Tropical	4	7	2	18	31
Antiboreal	8	0	9	21	38
Antarctic	1	1	7	5	14

Source: Data from Dogiel (1964).

Seventeen species were listed by Dogiel (1964) as occurring in pinnipeds and cetaceans, of which eight were found in the northern hemisphere only, five in the southern hemisphere only and four in both hemispheres. Helminth communities of freshwater fish in the high Arctic harbour several species of digeneans and cestodes, but acanthocephalans were notably absent from arctic char *Salvelinus alpinus* from lakes in north Norway and the islands of Svalbard (Kennedy, 1977, 1978a), and in this region the most likely explanation was the absence of suitable intermediate host species. Dogiel (1964) also suggested that the absence of acanthocephalans from fish in the Aral Sea related to the absence of suitable species of intermediate host.

Detailed studies from Mexico confirm this pattern. Vidal-Martinez & Kennedy (2000a) reported only four species of acanthocephalans from 40 species of freshwater fish from south-eastern and central Mexico. Two of these, *Acanthocephalus dirus* and *Octospinifer chandleri*, had been recorded from only one locality and from one species of host, and *Floridosentis mugilis* had only been recorded from three localities and four species of cichlid. Only one species could be considered widespread and fairly common: *Neoechinorhynchus golvani* had been found in 38 localities and occurred in 14 species of hosts, mainly species of cichlids. It occurred in all regions of Mexico except for the Rio Grande basin in the north-east, and seemed to be neotropical in its distribution rather than nearctic. Vidal-Martinez *et al.* (2001) in their review of helminths of Mexican cichlids came to a similar conclusion. Although they reported seven species of acanthocephalans from cichlids, only five occurred as adults in the intestine and only *N. golvani* was widely distributed. Acanthocephala were also poorly represented in fish from central America (Perez-Ponce de Leon & Choudhury, 2005). In contrast, 51 species have been recorded from the freshwater fish of the USA (Hoffman, 1999) and 32 species from fish in Canada (Arai, 1989).

A number of detailed studies have been conducted into the parasites of cichlid fishes in Mexico in particular. Salgado-Maldonado & Kennedy (1997) studied helminth community richness in *Cichlasoma urophthalmus* in seven localities in the Yucatan peninsula. The total number of species in each locality ranged from 9 to 23, but only a single species of acanthocephalan, *Neoechinorhynchus golvani*, was found in four of the lagoons and that always formed a very low proportion of the number of helminth individuals in the sample and generally occurred at a very low intensity (Table 3.6). Data from other species of cichlid from other lakes in another region of Mexico collected by Pineda-Lopez (1994) showed a very similar pattern. Once again *N. golvani* was the only species of acanthocephalan found, whereas, by contrast, in eels *Anguilla rostrata* in temperate North America, for example, 10 species of acanthocephalan can be found (Table 3.7). Although *N. golvani* occurred at higher levels of prevalence and intensity than in Yucatan (Table 3.8), it never dominated the helminth community in the way that a single species of acanthocephalan may do in fish in northern Europe, for example in *Pomphorhynchus laevis* in *Barbus barbus* (Table 3.9.). These data thus confirm the above suggestion that acanthocephalan communities are poorer in tropical regions than in temperate ones.

Table 3.6. *Occurrence of* Neoechinorhynchus golvani *in* Cichlostoma urophthalmus *in Mexican lagoons in the Yucatan peninsula*

	Lagoon						
	El Vapor	Palizada	Pargos	Cayo	Celestun	Chelem	Lagartos
p_i	0.002	0.09	0	0.01	0.005	0	0
Intensity	2.05	22.68	0	0.95	2.8	0	0
No. ex.	85	19	28	42	120	30	26
No. spp.	23	13	7	9	15	11	11
No. ps	1027	252	19	95	5611	134	1998

p_i, relative abundance of *N. golvani* expressed as a proportion of the total number of all intestinal helminths of all species; intensity, mean intensity of *N. golvani*; No. ex., number of fish examined; No. spp., number of helminth species found; No. ps, total number of intestinal helminths.
Source: Data from Salgado-Maldonado & Kennedy (1997).

Table 3.7. *Species of Acanthocephala recorded from species of eel*

North America	Europe	Australia and New Zealand
Host		
Anguilla rostrata	Anguilla anguilla	Anguilla reinhardtii
		Anguilla australis
		Anguilla dieffenbachii
Source		
Arai (1989), McDonald &	Kennedy (1990)	Kennedy (1995),
Margolis (1995),		Hewitt & Hine (1972)
Marcogliese & Cone		
(1998), Hoffman (1999)		
Acanthocephala		
Echinorhynchus salmonis	Echinorhynchus truttae	Telosentis australiensis
E. lateralis	Telosentis exiguus	Acanthocephalus galaxii
E. coregoni	Corynosoma semerme	Neoechinorhynchus aldrichettae
Acanthocephalus dirus	Acanthocephalus clavula	
Leptorhynchoides thecatus	A. anguillae	
Fessisentis friedi	A. lucii	
Pomphorhynchus bulbocolli	Pomphorhynchus laevis	
Neoechinorhynchus cylindratus	Neoechinorhynchus rutili	
Paratenuisentis ambiguus	Paratenuisentis ambiguus[a]	
Tanaorhamphus ambiguus		

[a]Species introduced into Europe.

Other patterns in acanthocephalan distributions across the globe are apparent. Acanthocephalan species richness is far greater in North America and Europe than in Australia and New Zealand. This is the case when a single species of host, for example eel (Table 3.7), forms the basis of comparison. It is also true when all species of fish are considered: Arai (1989) reported 32 species of acanthocephalan from Canadian waters, whereas Hewitt & Hine (1972) reported only four species from New Zealand. Three of these species were endemic to New Zealand and one of them used the introduced trout *Salmo trutta* as a paratenic and accidental definitive host.

Table 3.8. *The occurrence of* Neoechinorhynchus golvani *in cichlids in four lakes in Mexico*

	Species of cichlid											
	Cichlostoma helleri			*Cichlostoma synspillum*			*Cichlostoma urophthalmus*			*Petenia splendida*		
Lake	1	2	3	1	2	3	1	2	3	1	2	3
Prevalence	0	19.0	3.4	76.1	48.6	8.1	3.3	9.4	12.0	20.0	6.6	0
Mean int.	0	3.7	1.0	12.6	9.6	1.0	1.0	3.5	2.7	2.0	2.5	0
SD	0	4.7	0	17.5	11.5	1.2	0	0.3	3.6	1.0	1.5	0
N	100	269	87	535	607	37	30	127	343	15	30	25

Mean int., mean intensity; SD, standard deviation; N, number of fish examined.
Source: Data from Pineda-Lopez (1994).

Table 3.9. *The occurrence of* Pomphorhynchus laevis *in barbel* Barbus barbus *in some localities in Europe*

	Locality				
	R. Stour	R. Severn	R. Tiber	R. Rhone	R. Danube
Prevalence	100.0	100.0	80.0	100.0	100.0
Mean int.	169.0	56.2	14.0	44.6	71.0
Max. int.	445	643	121	150	491
Source	Kennedy (unpublished)	Brown (1989)	De Liberato et al. (2002)	Van Maren (1979b)	Moravec et al. (1997)

int., intensity.

Other endemic acanthocephalans have been recorded, including *Metechinorhynchus salmonis baicalensis* from Lake Baikal (Baldanova & Pronin, 2001) and *Pomphorhynchus patagonicus* from endemic species of galaxid fish in Patagonia, Argentina (Ortubay et al., 1991). Relict hosts are not, however, of necessity parasitised by relict parasite species, as seals of the Caspian Sea and Lake Ladoga are not parasitised by endemic or relict species of acanthocephalans (Dogiel, 1964).

As mentioned earlier, care must be taken in analysing distribution patterns due to uneven sampling of some host species and some regions of the world. Furthermore, the frequent movement of hosts around the globe for stocking, farming and other similar purposes may completely disrupt natural patterns of occurrence.

Despite this. it is still possible to recognise some patterns. The genus *Longicollum*, for example, contains two species: *L. pagrosomi* appears to be restricted to Japan's Inland Sea and the Sea of Japan, and *L. aleminsicus* is known only from Japan, Taiwan and Queensland, Australia (Yamaguti, 1963; Roubal, 1993). The closely related species *Tenuiproboscis misugurni* is also known only (Yamaguti, 1963) from the Inland Sea of Japan from the stomach of the fish *Misugurnis fossilis*: the report of a congeneric species from a stray dog in Algeria does strongly suggest a misidentification! The distribution of the freshwater species *Echinorhynchus salmonis* in Europe is closely linked to the distribution of its intermediate host, the glacial relict crustacean *Pontoporeia affinis* (Dogiel, 1964). Because this is the only known intermediate host for this species, its distribution is restricted to the occurrence of its intermediate, rather than definitive, host; so it is possible to be certain that a record of it from the British Isles is a misidentification since the intermediate host does not occur there (Chubb, 2004).

3.2.2   Regional patterns

The distribution of the genus *Acanthocephalus* is worldwide, but whereas some species such as *A. lucii* and *A. clavula* are widespread throughout the continent of Europe, others appear to be very local, for example *A. acerbus* and *A. aculeatus* in Japan and *A. amuriensis* in the Amur basin. However, these apparently local distributions may well reflect a lack of widespread sampling. This is not the case in North America, where Amin (1984, 1985b, 1986) has made a detailed study and analysis of the interspecific and intraspecific variation in the genus. As a consequence, Amin (1985b) recognised three species in North America. The species *A. dirus* has the widest distribution range, occurring throughout the Mississippi Basin and places such as Lake Erie that were once linked to it. The other two species can be characterised as southern and more local in distribution (Fig. 3.2). *Acanthocephalus alabamensis* occurs only in Alabama in six species of hosts belonging to four families and only in localities formerly linked to the Mississippi basin. The third species, *A. tahlequahwensis*, is found only in Oklahoma in four host species belonging to two families and only in tributaries of the Mississippi. The three species are believed to be monophyletic (Amin, 1986), with *A. dirus* being the ancestral species. The two southern species are closer to each other than to *A. dirus*: they evolved allopatrically and now show restricted

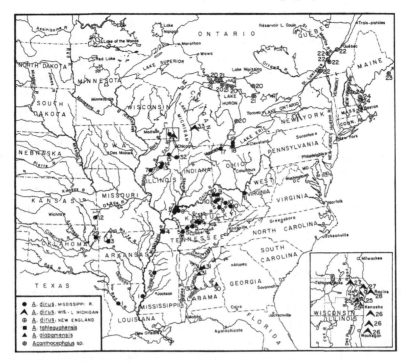

Fig. 3.2 Map of the eastern United States showing the present
distribution of *Acanthocephalus* spp. from freshwater fish in the
Mississippi River drainage system, Great Lakes and the St Lawrence
watershed as well as from New England, Nova Scotia and Alabama.
(From Amin, 1985b, with permission of the author.)

hosts and distributions. However, even within *A. dirus* there are distinct subgroups. Amin believes that *A. dirus*, the most morphologically
variable species, was originally widespread in the Mississippi river
before the Wisconsin glaciation. It then spread to the Great Lakes at
the time of the retreat of the glaciers and its current distribution is
linked to that of its intermediate host, the isopod *Asellus intermedius*.
In Wisconsin, the Lake Michigan population was isolated post-glacially,
c. 15 000 years BP. This population uses *Pontoporeia affinis* as its intermediate host and shows some variation, perhaps incipient speciation,
since isolation. A river population uses *Caecidotea militaris* as intermediate host. Finally, he recognises a New England population, not
linked to the St Lawrence in any way or to the Mississippi, even though
its morphology suggests a Mississippi source.

The genus and species of *Pomphorhynchus* also show both global and local patterns in distribution. Although 27 species have been reported from the genus, according to Brown (1987) six of these exhibit characteristics that fall within the range of a single population of *P. laevis*. Several other species are either poorly described, or have been reported on one occasion only and so are regarded as of doubtful validity. This leaves ten species that currently appear to be valid (although some of the doubtful ones from China and India will probably turn out to be so), and these show quite discrete patterns of distribution and host usage (Table 3.10). Half of them have been reported from fresh water, and half from euryhaline or sea water. Some are very local in distribution, for example, *P. heronensis* has only been found on the Great Barrier Reef in Australia (Pichelin, 1997), *P. patagonicus* from Patagonia (Ortubay *et al.*, 1991) and *P. yamagutii* from Chile (Schmidt & Hugghins, 1973). It appears therefore that *Pomphorhynchus* is an old genus with a global distribution, but species have evolved in discrete zoogeographical regions to use local fish and amphipod species as hosts and so they differ biologically in their host and habitat requirements (Kennedy, 2003).

It was originally believed that the European species *P. laevis* was both widespread throughout Europe and widely specific to its intermediate and definitive hosts. The current belief, however, is that there are three strains within Europe, each of which uses a different definitive and intermediate host (Table 3.10). Two of these strains are freshwater and the third is marine. Throughout the continent, the species uses *Barbus barbus* and *Leuciscus cephalus* as definitive hosts and the local species of freshwater *Gammarus* as intermediate host. This is believed to be the original, post-glacial pattern of host usage (Kennedy *et al.*, 1989). When the Thames–Rhine basin broke up with the formation of the North Sea, this remained the pattern of host usage in the Thames and the Rhine. In the Baltic and North Sea, a strain evolved that used marine flounders, *Platichthys flesus*, as definitive host and marine species of *Gammarus* as intermediate host. The subsequent distribution of *P. laevis* in England reflects the anthropogenic stocking of *Barbus barbus* into some river systems from the River Thames (Fig. 3.3a). The situation in Ireland differs again, as here the species uses *Salmo trutta* as its definitive host and *G. duebeni* as intermediate host. It has been suggested that *P. laevis* was introduced into Ireland from England with cyprinid fish by humans (Kennedy *et al.*, 1989). In the absence

Table 3.10. *Currently recognised valid species of* Pomphorhynchus

	Definitive host	Intermediate host	Distribution
Palaearctic			
P. laevis[a]	Barbus barbus	Gammarus pulex	England, European continent
	Leuciscus cephalus	Gammarus fossarum	
	Salmo trutta	G. duebeni	Ireland, Scotland
	Platichthys flesus	G. zadachi	Baltic North Sea
		G. locusta	
P. perforator	Diptychus dybowskii	Gammarus bergi	Turkestan, China
	Schizothorax pseudooxainsis	G. lacustris	
Nearctic			
P. bulbocolli	Micropterus spp.	Hyalella azteca	North America: Canada & USA
	Ictalurus punctatus + 38 other species	Gammarus tigrinus G. pseudolimnaeus	
P. lucyi[a]	Erimyzon sucetta +14 other species	unknown	SE USA
P. rocci[a]	Micropterus salmoides	Gammarus tigrinus	USA
Neotropical			
P. patagonicus	Galaxias platei Patagonina hatcheri Perchichthys trucha	Hyalella patagonica	Argentina, Patagonia
P. sphaericus	Pimelodes maculatus P. albicans	Hyalella sp.	Argentina
P. yamagutii	Percichthys melanops	unknown	Chile
Australia			
P. heronensis[a]	Lujanus carponotatus	unknown	Queensland
Oriental			
P. sebastichthydis	Sebastichthys oblongatus	unknown	Japan

[a]Also from euryhaline or sea water; all others fresh water.
Source: From Kennedy (2003), with permission, Perpignan University Press.

of its preferred definitive hosts there it used its known ability to reproduce in *Salmo trutta*, evident in rivers in northern Italy (Dezfuli *et al.*, 2001a), to survive in Ireland and has subsequently adapted to this host there.

(a)   (b)

Fig. 3.3 Maps of the British Isles showing the distribution of
(a) *Pomphorhynchus laevis* (circles, freshwater strain; squares,
marine strain) and (b) *Acanthocephalus anguillae*. (From Kennedy
et al., 1989, with permission.)

The species *Acanthocephalus anguillae* also uses *L. cephalus* as a
preferred definitive host but *Asellus aquaticus* as its intermediate
host, and is restricted to river systems in eastern England that
were part of the Thames–Rhine basin (Fig. 3.3b). On introduction
to Ireland, this species used its ability to reproduce in eels to
survive and now uses this as its normal definitive host there
(Kennedy *et al.*, 1989). The exclusive distributions of *P. laevis* and
A. anguillae in Britain are also believed to reflect the fact that
these two species compete in their definitive host (pp. 152–3) and
utilise different species of intermediate hosts. Thus, in three
genera at least, and in Europe and North America, the pattern
of host usage by species still reflects post-glacial events. In other
genera, patterns are completely different: in *Neoechinorhynchus*,
for example, the distribution of some species is linked to the
distribution of freshwater turtles, or cichlid fish, or, as in the case
of *N. rutili*, it is widespread through North America, Europe and
parts of Asia and infects fish belonging to at least 37 different
genera.

3.3 HOST FEATURES

3.3.1 Aquatic hosts

Although adult acanthocephalans have been reported from a wide variety of vertebrate definitive host species, even a cursory survey of the literature suggests that some groups of hosts and some host species may be more frequently parasitised than others that are only rarely infected or only infected accidentally. However, interpretation of such data presents major problems as many species have only been studied on a few occasions, sampling has been very uneven in respect of time and place, and negative records are not always published. Nevertheless, even with these reservations, it is possible to detect a pattern in respect of those species that are, and perhaps more interestingly those species that are not, regularly parasitised by acanthocephalans.

Freshwater species

Freshwater fish, and especially those with an omnivorous or pis-civorous diet, are frequently to be found harbouring acanthocephalan species (Bangham, 1955; Arai, 1989; Kennedy, 1990; Williams & Jones, 1994; Kennedy & Hartvigsen, 2000). Moreover, in some hosts and localities, acanthocephalans may be the dominant intestinal parasites. In a survey of the parasites of eels *Anguilla anguilla* in 50 localities in the British Isles, Kennedy (1990) found that 54.7% of them were dominated by a species of acanthocephalan. This situation is common in eels in Ireland (Conneely & McCarthy, 1984, 1986; Callaghan & McCarthy, 1996) and in continental Europe (Moravec, 1985; Schabuss *et al.*, 1997; Kennedy *et al.*, 1998; Sures *et al.*, 1999a). Acanthocephalans may dominate in some inland European lakes which are stocked with eels, such as Neusiedler See (Schabuss *et al.*, 2005), but may be absent from others such as Lake Balaton (Molnar & Szekely 1995). Acanthocephalans are still present in eels from freshwater lagoons around the Mediterranean (Kennedy *et al.* 1997) but disappear as salinity levels rise (Di Cave *et al.*, (2001). In Portugal, where species of *Asellus* that serve as intermediate hosts for *Acanthocephalus* species are absent or very local in occurrence, acanthocephalans are present in eels but rarely dominant (Saraiva & Eiras, 1996). In North America, acanthocephalans do occur in populations of *Anguilla rostrata*, but seldom dominate (Cone *et al.*, 1993;

Marcogliese & Cone, 1993). In New Zealand, acanthocephalans are widespread but intermittent in their occurrence in eels (Hine, 1978) and in Australia only one species occurred in eels, and that rarely, but never dominated (Kennedy, 1995).

In Europe, helminth communities in other species of fish may also be dominated by an acanthocephalan species: in barbel *Barbus barbus* in particular, *Pomphorhynchus laevis* is frequently the dominant species (Hine & Kennedy, 1974a; Van Maren, 1979a,b; Moravec *et al.*, 1997; Laimgruber *et al.*, 2005) and in the River Tiber it is subdominant to *Acanthocephalus clavula* (De Liberato *et al.*, 2002). Acanthocephalans are also frequently to be found in helminth communities of brown trout *Salmo trutta*, and in the British Isles they dominated 55% of the 38 localities sampled (Kennedy & Hartvigsen, 2000). They are also widespread in cyprinid fish in many localities, for example in Lake Druzno (Styczynska, 1958). In Canada, the mean diversity (mean number of helminth species per host individual) is lower for acanthocephalans (0.258 compared with 0.379 for cestodes) and 0.65 for trematodes across a range of fish species (Bell & Burt, 1991), suggesting that they may be less dominant than in Europe. It seems that they are commoner and more likely to be dominant in fish with an omnivorous diet.

Some specialised feeders such as species of sturgeon *Acipenser* are frequently infected with acanthocephalans (Ibragimov, 1985; Moravec *et al.*, 1989), but others, such as lesser sand eels *Ammodytes tobianus*, appear to be free of acanthocephalans (O'Connell & Fires, 2004). Moreover, detritus feeders, such as *Cyprinus carpio*, or vegetarian species such as *Ctenopharyngodon idella*, seldom or never harbour acanthocephalans (Nie, 1995; Kennedy & Pojmanska, 1996) although they can be common in *Gymnocypris przewalskii przewalskii* in selected localities (Yang & Liao, 2001). In a recent study, Karvonen *et al.* (2005) have queried why Crucian carp (*Carassius carassius*) generally harbour depauperate parasite communities. When they compared the parasite fauna of Crucian carp in several localities where it occurs alone with other localities in which it occurs with other fish species, they found that the helminth fauna differed both qualitatively and quantitatively. When alone, the helminth fauna was species poor and no acanthocephalans were present. When the Crucian carp occurred with other species of fish, the helminth fauna was richer and in some localities (<50%) included acanthocephalans, namely *Acanthocephalus lucii* and *Neoechinorhynchus rutili*, albeit at low levels of prevalence and abundance. They considered that the depauperate parasite fauna of Crucian carp could reflect: the absence of other host species in a water body, which limits the possibility of parasite exchange; or the unsuitability of many habitats for the invertebrate species that serve as intermediate hosts; or the physiology and diet of this host species. They believed that this latter was the most important factor, although this did not completely prevent them from acquiring generalist acanthocephalans from other fish hosts in small numbers.

It is important to appreciate that acanthocephalans may be absent from a suitable host in some localities, for example from *Salmo trutta* in a group of four isolated lakes in Ireland (Byrne *et al.*, 2000), and from char *Salvelinus alpinus* in high arctic lakes (Kennedy, 1977, 1978a) and from several species of native and non-native fish in the Little Colorado River in Arizona (Choudhury *et al.*, 2004). The explanation in many of these cases may relate to the absence of suitable intermediate hosts from the localities. It has been noted earlier in Tables 3.6 and 3.8 that acanthocephalans are poorly represented amongst the parasites of Mexican cichlids and this seems to be equally true in the case of African cichlids (Aloo, 2002). It is also essential to realise that whereas one species in a genus may be uninfected by acanthocephalans, for example *Fundulus zebrinus* (Janovy & Hardin, 1988: Hoffman, 1999) other species in the same genus may serve as hosts to several species (Hoffman, 1999). Here again, presence and abundance of acanthocephalans seems to be related to host diet.

Marine species

Marine fish may often be hosts to acanthocephalans, although deep-water species and surface-feeding species are more likely to be infected than mid-water species (Buron, 1988; Holmes, 1990; Rohde, 1993; Buron & Morand, 2004), but it is more difficult to detect patterns in the data sets. This may reflect less extensive data sets and sampling problems, but it may also be the case that the occurrence of acanthocephalans in marine fish is far less predictable. In his data set of 88 individuals from 10 populations of *Sebastes nebulosus*, Holmes (1990) found only one species of acanthocephalan and that was never dominant. By contrast, *Rhadinorhynchus ganapatii* was quite common, even if never dominant, in a population of tuna (Madhavi & Sai Ram, 2000). A number of species have been recorded from cod, *Gadus morhua* (Polyanski, 1961; Huffman & Bullock, 1975; Arai, 1989), including *Echinorhynchus gadi* in a relict cod population (Kulachkova & Timofeeva, 1977), and from both Pacific halibut *Hippoglossus stenolepis* (Blaylock *et al.*, 1998) and Atlantic halibut *Reinhardtinus hippoglossoides* (Arthur & Albert, 1994), but many of these are larvae using the species as a paratenic host and the occurrence of adults is erratic. Fish in some localities, for example *Mugil*, *Liza* and *Siganus* species off the coast of Israel, harbour several species of acanthocephalans (Schmidt & Paperna, 1978; Diamant, 1989), but these may be the exception as elsewhere only one species of acanthocephalan was found in *Liza* and that was uncommon (Di Cave *et al.*, 1997). Frequently, no acanthocephalans at all are found (Holmes & Bartoli, 1993; Bartoli *et al.*, 2000) or no adults, but only encysted larvae in the mesenteries (Sanchez-Ramirez & Vidal-Martinez, 2002). In general, marine helminth communities of fish are dominated by species of digeneans, and the occurrence of acanthocephalans relates to the host diet.

Marine turtles have seldom been examined, but the study by Aznar *et al.* (1998) on a sample of *Caretta caretta* revealed no acanthocephalans. The authors believed that marine turtles had an association with their parasites largely independent of that of other marine hosts and that helminth communities were depauperate and dominated by digeneans. This contrasts markedly with the situation in freshwater turtles in which several species of acanthocephalans may co-occur (Aho *et al.*, 1992). Marine mammals, by contrast, harbour helminth communities that may be dominated by acanthocephalans (Balbuena & Raga, 1993; Aznar *et al.*, 1994): Balbuena & Raga reported that *Bolbosoma capitulum* may be the most prevalent species in long-finned pilot whales, although they are less abundant than digeneans. Helminth communities in seals are frequently dominated by acanthocephalan species (Helle & Valtonen, 1980; Valtonen & Helle, 1988; Valtonen & Crompton 1990; Nickol *et al.*, 2002b) and indeed these may be the only helminth species present in some populations (O'Neill & Whelan, 2002). It would again appear that the presence of acanthocephalans is related to diet – piscivorous mammals being able to acquire them from fish paratenic hosts.

3.3.2 Terrestrial hosts

Herptiles

Although there are some species of acanthocephalans that are specialists of amphibia and reptiles, for example *Acanthocephalus ranae* and *A. sinensis*, it appears that many herptiles are uninfected by acanthocephalans. The extensive review by Aho (1990) reported only one species from a wide range of amphibian and reptile hosts in North America and this was nowhere widespread or abundant. Amphibians harbour several species of nematodes and often therefore rich intestinal communities, but acanthocephalans are at best only a minor component and are frequently absent altogether from a locality (Muzzall, 1991; Bursey & Goldberg, 1998; Boquimpani-Freitas *et al.*, 2001). The same may be said of reptiles (Biserkov & Kostadinova, 1998; Matsuo & Oku, 2002; Martin & Roca, 2004; Menezes *et al.*, 2004), and all these authors agree that their helminth communities are dominated by direct life-cycle nematodes. Where acanthocephalans have been reported, for example by Roca & Hornero (1994) and Vrcibradic *et al.* (2002), the acanthocephalans are often in the cystacanth stage and so the host is acting as a paratenic host rather than a definitive one.

Birds

Checklists and systematic studies, such as Yamaguti (1963), suggest that birds are frequently parasitised by acanthocephalans, and it is known for example that acanthocephalans can be found in many species of owls and raptors. However, there are relatively few ecological studies from which their dominance, or lack of, can be ascertained. Hynes & Nicholas (1963), Thompson (1985a,b,c), Bush & Holmes (1986a,b), Goater & Bush, (1988), Goater (1989) and Bush (1990) have analysed parasite communities in populations of ducks and waders. In many cases the helminth communities were very rich and speciose, and in all cases included acanthocephalan species, but often these occurred at low prevalence and intensity levels and did not dominate the communities. The exception to this came in the studies by Liat & Pike (1980) and Thompson (1985a,b,c) on eider ducks, but the acanthocephalans in this case appeared to have reached unusual epizootic levels. Edwards & Bush (1989) studied the parasites of avocets in four localities, and found only one species of acanthocephalan. This occurred at low levels in three localities, but in one lake, where amphipods were particularly abundant, it occurred at very high levels of prevalence and intensity. The studies by Stock & Holmes (1988) on four species of grebes also showed that acanthocephalans were present, but the communities were dominated by cestodes. Many other studies of bird parasites over a range of hosts including starlings (Owen & Pemberton, 1962; Moore 1983a), swallows (Sherwin & Schmidt, 1988), woodpeckers (Nickol, 1977; Foster *et al.*, 2002), bobwhite quail (Moore & Simberloff, 1990), oystercatchers (Goater, 1989) and even bustards (Jones *et al.*, 1996) have demonstrated the presence of one or more acanthocephalan species, but at low levels of infection. Species that scavenge, such as herring gulls, may not harbour acanthocephalans (Threlfall, 1967), nor may plant feeders such as the red-legged partridge (Calvete *et al.*, 2004) or Barbary partridge (Foronda *et al.*, 2005). Overall the presence of acantho-cephalans in birds seems to relate closely to their diet: aquatic birds feeding on arthropod intermediate hosts may aquire them directly, but in the case of more terrestrial birds, and especially raptors, they probably aquire them by feeding on paratenic vertebrate hosts. Whatever the method of aquisition, acanthocephalans seldom seem to dominate the helminth communities.

Mammals

As is the case with birds, it is difficult to form an opinion on the occurrence and importance of acanthocephalans in helminth com-munities of mammals. Pence (1990) in his review focused on three species, albeit from several localities in each case. White-tailed deer did not harbour any acanthocephalans and communities were dominated by nematodes; although black bears did act as hosts to one species of acanthocephalan, this occurred only at low infection levels and helminth communities were dominated by nematodes. Two species of acanthocephalan were reported from coyotes but from only two localities and neither was ever dominant. Other studies tend to confirm this pattern. Studies on parasites of goats in Nigeria (Nwosu *et al.*, 1996), antelope in South Africa (Fellis *et al.*, 2003), bovids in Kenya (Ezenwa, 2003) and a long-term study of ponies in Louisiana (Chapman *et al.*, 2002), confirm that acanthocephalans are normally absent from ungulates, the communities of which are dominated by nematodes. Acanthocephalans can be found in carnivores, but only in some species and some places, and their occurrence is rather erratic (Pence & Windberg, 1984; Papadopoulos *et al.*, 1997; Moro *et al.*, 1998) and in some cases they may be absent altogether (Bussche *et al.*, 1987; Torres *et al.*, 1998; Segovia *et al.*, 2001). They are very uncommon in wolves (Craig & Craig, 2005) but have been reported from urban foxes in Copenhagen (Willingham *et al.*, 1996) and, albeit rarely, in otters (Torres *et al.*, 2004). One species, *Macracanthorhynchus ingens*, is a specialist of racoons, where it is common in some localities (Bafundo *et al.*, 1980; Smith *et al.*, 1985), although never dominant, but rare in others (Snyder & Fitzgerald, 1985).

Acanthocephalans appear to be absent from hares (Hurlbert & Boag, 2001) and rabbits (Allan *et al.*, 1999; Foronda *et al.*, 2003). They are also absent from squirrels (Patrick, 1991), wood mice (Fuentes *et al.*, 2004), spiny mice (Behnke *et al.*, 2000) and voles (Abu-Mahdi *et al.*, 2000; Behnke *et al.*, 2001; Barnard *et al.*, 2003). They may be absent from rats in some localities (Abu-Mahdi *et al.*, 2001, 2005), but *Moniliformis moniliformis* may dominate the helminth community in others (Mafiana *et al.*, 1997). They may similarly be absent from pigs in some localities (Boes *et al.*, 2000) but be present and dominate helminth communities in wild boars in others (Solaymani-Mohammadi *et al.*, 2003). They are often absent from marsupials (Griffith *et al.*, 2000), or at best very uncommon (Table 3.11) in comparison with nematodes (Beveridge & Spratt, 1996). They have been reported as being uncommon in tarsiers and as an accidental infection in humans (Schmidt, 1971). They have not been recorded from bats (Lotz & Font, 1994).

Overall, the occurrence of acanthocephalans in terrestrial vertebrates seems to relate closely to diet. Strict herbivores do not normally harbour acanthocephalans, and they are uncommon in strict carnivores. They are more likely to be found in species with a more omnivorous diet which includes invertebrates, or in some predators which ingest species that serve as paratenic hosts. Seldom, if ever, however, do they dominate the helminth communities in these hosts.

Infections of humans

There are records of acanthocephalan infections from humans, but these are erratic and spasmodic and seem to represent accidental

Table 3.11. *Parasites of Australian marsupials*

	Trematodes	Cestodes	Nematodes	Acanthocephlans	Total
No. of families	9	5	24	2	40
No. of endemic families	0	0	5	0	5
No. of genera	14	16	99	2	131
No. of species	26	77	327	2	432

Source: Modified from Beveridge & Spratt (1996), with permission from Elsevier.

Table 3.12. *Records of acanthocephalans from humans*

Species	No. of Records	Country	Status	Source of infection
Acanthocephalus bufonis	1	Indonesia	A	?
Acanthocephalus rauschi	1	Alaska	A	fish
Bolbosoma sp.	2	Japan	A	fish
Corynosoma strumosum	2	Alaska	A	fish/seal
Macracanthorhynchus hirudinaceus	5	China, USSR, Thailand, Madagascar	A	insects
Macracanthorhynchus ingens	1	Texas	A	insects
Moniliformis moniliformis	Several	9 countries	A	insects

A, accidental and dead end as does not become sexually mature. The source is always presumed.
Source: Data from Coombs & Crompton (1991) and Ashford & Crewe (1998).

infections of no ecological significance since the parasites do not mature in humans and cannot transmit infections from this source (Table 3.12). Most of these infections can be considered to result from dietary 'mistakes', either as a result of eating raw fish through choice (the infections of individual Inuits in Alaska with *Acanthocephalus rauschi* and *Corynosoma strumosum* and of Japanese with *Bolbosoma* sp.) or of eating insects for whatever reason (infections with *Macracanthorhynchus* species and *Moniliformis moniliformis*). Only this latter species has been reported from humans on more than a handful of occasions and over a wide geographical range and clearly reflects the close contact between rats and humans. No species is specific to humans or occurs regularly in them: occurrences seem to reflect particular localised dietary preferences or aberrations.

In so far, therefore, as any pattern is discernible in the occurrence of acanthocephalans in vertebrates, it can be summed up as 'you get what you eat'. Because acanthocephalans can only be transmitted to their definitive hosts indirectly by ingestion of infected arthropod intermediate hosts, or by ingestion of vertebrate paratenic hosts, the presence of adults in a vertebrate will relate closely to the diet of that species. Only species that are omnivorous, piscivorous or carnivorous in their diets are likely to be regularly infected with acanthocephalans. Superimposed on this are local patterns, in that not all populations of an otherwise susceptible species are likely to be infected since intermediate hosts may be absent from any locality, and global patterns, in that acanthocephalans are less common in tropical regions and on land. In general, though, herbivores and grazers, whether fish, reptiles, birds or mammals, are unlikely to become infected, as are seed-eating birds and rodents. Dominance of helminth communities by acanthocephalans appears to be more characteristic of temperate freshwater fish and aquatic mammals. Parasite communities in marine fish are generally dominated by digeneans and in herptiles: ungulates and other grazing mammals by nematodes. Reliance on an intermediate host does limit the distribution of acanthocephalans: a limitation that the nematodes have overcome by employing direct life cycles and infecting grazing herbivores.

4

Specificity

4.1 GENERAL STRATEGY

The passage of an acanthocephalan through its life cycle involves ingestion by a host at each stage, from egg to intermediate host, and from intermediate host to paratenic or definitive host. The parasite moves up the trophic levels in a series of stages (and down in one by eggs) and through a food web (Marcogliese, 2002). Although the choice of a particular host will tend to canalise the direction of movement, it is highly improbable that an egg or intermediate host will be eaten *only* by a species that will serve as the preferred intermediate or definitive host: both eggs and intermediate hosts will be eaten by a number of other species. Some of these will be the preferred host in which the parasite will grow and develop and with which it will form a stable, balanced system. In other species the acanthocephalan may be able to develop, but less successfully and/or more slowly, and these species can be considered to be suitable but secondary hosts. In a third group of hosts the acanthocephalan may be able to survive, but be unable to grow or develop (or only very occasionally) and these can be considered to be accidental or unsuitable hosts. Finally, in many species it will be unable to survive at all and will die rapidly, for example if *Macracanthorhynchus ingens*, a parasite of racoons *Procyon lotor*, accidently infects pigs it survives for only two weeks (Nelson & Nickol, 1986). This ability of a parasite to form a spectrum of relationships with host species is referred to as its specificity. If it is able to use only one or a very few (often closely related) species of host, the acanthocephalan is considered to be very host specific: if it can use many and unrelated hosts, it is considered to be widely specific.

The specificity of any particular parasite can be viewed as a compromise between the two extremes of specialisation, i.e. adaptation to only a single species of host, and generalisation, i.e. the ability to utilise a wide range of host species. As with all animal species, some degree of specialisation is normal as a species cannot adapt to everything. Narrow specificity may often be accompanied by heavy parasite mortality as a consequence of parasites being ingested by hosts in which they cannot survive. Overspecialisation of course carries with it the danger that if the host population or species becomes extinct, then so does the parasite, but a high degree of adaptation may confer a competitive advantage on the parasite. Although one might expect most species to adopt an intermediate position as their evolutionary strategy, thereby hedging their bets by achieving a trade-off between specialisation and flexibility, this is seldom the case. In any class of parasites, some species will be narrowly specific and some more widely specific and even congeners may adopt quite different strategies, for example species of *Mediorhynchus* and *Macracanthorhynchus* (Table 2.3). Moreover, acanthocephalans may show different degrees of specificity to their intermediate, paratenic and definitive hosts (Table 2.3) and be narrowly specific to one host and widely specific to another.

The majority of acanthocephalan life cycles are still unknown. Golvan (1957) considered that acanthocephalans were more specific to their definitive hosts than to their intermediate and paratenic hosts, but stressed that the basis for this generalisation was weak and based largely on very few species. While this might be the case for some terrestrial species such as *Moniliformis moniliformis* and *Macracanthorhynchus hirudinaceus*, it is not true for others, for example *Moniliformis clarki* (Table 2.3). Golvan correctly recognised that there was considerable variation between species, and that many acantho-cephalan species could attach and survive in a host species without reaching sexual maturity. Chubb (1967) demonstrated that specificity towards both intermediate and definitive hosts was variable, but could be strict in some cases. He stressed the need for more ecological and experimental investigations of specificity and concluded that a greater degree of host specificity might exist within the Acanthocephala than might be thought from a study of the literature then available. Twenty years later, Schmidt (1985) in his review of acanthocephalan life cycles could still provide information on only 125 species out of the estimated total of *c.* 1000 in the phylum. Thus, the basis for generalisations still remains smaller than one would like, especially when heterogeneity is viewed across the whole spectrum of genetic differences between host species, populations and individuals, and when founder effects and population isolation are also considered (Combes, 2001).

Table 4.1. *Host specificity of helminth parasites of fish in the Barents Sea*
Host records from other seas are considered: records of accidental hosts are not

	N.sp.	Percentage of species					
		In one host species	In one host genus	In one host family	Mainly in one family	In several families	Un.
Monogenea	21	52.4	9.5	33.3	4.8	0	0
Digenea	37	2.8	11.1	25.0	16.7	44.4	2.8
Cestoda	19	12.5	6.2	18.7	25.0	31.4	6.2
Nematoda	12	9.1	0	36.3	9.1	36.4	9.1
Acanthocephala	3	33.3	33.3	33.3	0	0	0

N.sp., number of species: Un., undetermined.
Data from Polyanski (1966).

Despite these reservations, there is a widely held view that acanthocephalans as a whole are more widely specific to their definitive host than most other groups of parasites, and this view is based on data sets such as that in Table 4.1. Despite the small number of acanthocephalan species, they do appear to be more widely specific to their fish definitive hosts than the Monogenea. This would be consistent with the concept of filters or screens to infection. Holmes (1987) considered that all parasites, including acanthocephalans, had to pass through a series of filters before they could establish a relationship with a host (Table 4.2). Some of these filters were environmental, e.g. zoogeographical factors and interactions with other species, and some were physiological, e.g. nutrition, diet and immunity of hosts. They can in fact be grouped and visualised as encounter factors, which determine whether an acanthocephalan will actually encounter a potentially suitable host, and compatibility factors, which determine whether it will survive if it does (Fig. 4.1). It is important to notice that some compatible species may never normally be encountered and the inherent danger of introducing species to a new region is that such encounters may then become possible (see Chapter 8). Any behavioural changes that a parasite may induce in its host (acanthocephalans almost invariably induce such changes in their intermediate hosts; Chapter 5) is also likely to widen the encounter window (Combes, 2001). Acanthocephalans also very often use key species in a community as their intermediate hosts, and these will be fed upon by a number of

Table 4.2. *Screens to parasitic infections of hosts*

Effect	Screen	Potential parasite fauna
Availability	Zoogeographical factors	
	Regional historical factors	Regional parasite fauna
	Local historical factors (natural disturbances)	
	Environmental factors (host community)	Local parasite fauna (Compound community[a])
	Anthropogenic factors	Observed parasite fauna (Compound community[b])
Contact	Food habits/exposure factors	
	Phylogenetic specificity	
	Physiological factors of species	
	Interactions between species	Species' parasite fauna (Component community)
Suitability and resistance	Physiological factors (individual characteristics: nutrition, stress, immunity)	Individual's parasite fauna (infracommunity)

[a]The Compound community we think we study as opposed to:
[b]the Compound community we actually study.
Source: Based on ideas presented by Holmes, (1987).

species at the next trophic level, and so the encounter window will be further widened. If, therefore, cystacanths are likely to encounter a wide range of predatory species as a consequence of food web structure (Marcogliese, 2002), the ability to survive in them and use them as a paratenic host or to develop to maturity in them as a definitive host would seem to be advantageous and so wider specificity may be a good evolutionary strategy. Clearly parasite specificity will have a major influence on the distribution and abundance of an acanthocephalan species. It follows from this that the hosts of acanthocephalans are more likely to be related ecologically through similar diets and habitat preferences than phylogenetically through a common ancestry. As Poulin (2005) has put it, 'ecological similarity amongst host species arising from convergence and not from relatedness is more important than host taxonomy'.

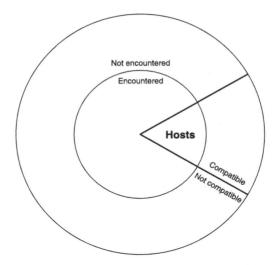

Fig. 4.1 Schematic representation of the forces determining the range of host species used by a parasite. Only a subset of all potential host species (outer circle) are actually encountered by an acanthocephalan (smaller inner circle represents the encounter filter). The triangular area represents the compatibility filter, which excludes all species in which the parasite cannot develop. Only species passing through both filters can become suitable hosts. Many compatible species may never actually be encountered. (Modified from Combes, 1991, by Poulin, 1998, fig 3.7, with kind permission of Springer Science and Business Media.)

4.2 CRITERIA FOR SPECIFICITY

A checklist of parasite species and their hosts such as those of Yamaguti (1963), Arai (1989) and Hoffman (1999) can, under some circumstances, provide a good indication of a parasite's specificity. However, its value as an indicator depends to a great extent upon two things: the parasite being narrowly specific and the number of records of a species. A mere list of hosts from which an acanthocephalan species has been recorded fails to distinguish the status of these hosts: in particular whether they are paratenic or definitive hosts and if the latter, whether preferred, suitable or accidental hosts. Such information is vital to determine the specificity of a species but in many cases it is not recorded because it is not known. It is also very important to have some measure of sampling effort. If a species has only been recorded once and then from a single species of host, then nothing is actually known about its specificity and no presumption should be made. Poulin (1992) and Poulin & Mouillot (2003) have shown that there is a very significant correlation between the number of known host species and the number of published records across all helminth species of freshwater fishes of Canada (Fig. 4.2). Considering only the acanthocephalans, just under 20% of the species had been recorded from only a single host species: the remainder had been reported from between 5 and 20 host species. There was a mean of 7.78 records per species and the mean number of hosts per species was 12.61 (Table 4.3). Although it was apparent that the acanthocephalans

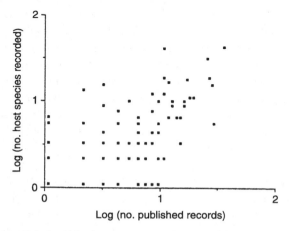

Fig. 4.2 Relationships between study intensity and the number of known hosts among 176 species of freshwater fish parasites. The regression between the two variables is highly significant ($r^2 = 0.50$, $P < 0.001$). (Reprinted from Poulin, 1992, copyright 1992, with permission from Elsevier.)

Table 4.3. *Comparisons of study effort and three measures of host specificity (means ± SD) among four taxa of helminths parasitising Canadian freshwater fishes*

Parasite taxon	N	No. of records/species	No. of host species	S_{TD}	$VarS_{TD}$
Trematodes	63	4.76 (5.03)	4.16 (4.58)	2.69 (1.03)	0.69 (0.46)
Cestodes	50	6.92 (8.52)	4.01 (4.21)	2.66 (0.94)	0.87 (0.73)
Nematodes	39	5.49 (5.45)	6.85 (9.12)	2.82 (0.84)	0.82 (0.56)
Acanthocephalans	18	7.78 (6.26)	12.61 (14.41)	3.05 (0.92)	0.94 (0.53)

N, number of species: S_{TD}, index of specificity.
Source: Modified from Poulin & Mouillot, (2003).

infected significantly more host species than other helminth groups, no distinction was made between the categories of infected hosts.

In a more recent study, Poulin & Mouillot (2005a) have also pointed out that different animal species have different probabilities of being discovered, related to a large extent to the biological character-istics of the species. Thus, species with broader geographical ranges are more likely to be encountered by scientists than species with restricted distributions. Applying this to parasites, they suggest that host specificity may similarly be linked to the probability of a host species being found. Using the same data set on helminth species of freshwater fishes in Canada as in Fig. 4.2, they showed that host specificity is associated with the year in which the helminths were described. Helminths that exploit more host species and/or a broader taxonomic range of host species tended to be discovered earlier than the more host-specific species. This relationship was apparent in all parasitic groups including the Acanthocephala. In one sense their findings represent a truism, as the longer the period over which a parasite has been known the greater is the probability that its range of hosts will also be known. There will also inevitably be exceptions, in that if a host species becomes of commercial interest for whatever reason, attention will inevitably be focused on its parasites. Nevertheless, their analysis serves as a valuable reminder that in many cases it may be difficult to evaluate the specificity of a parasite if such temporal relationships are not taken into consideration, as the estimates of a species' specificity relates to one moment in time and may well change over time.

In order to overcome this difficulty, there have been several attempts to devise an index of specificity. One of the most recent of these, with an ecological perspective, is that of Rohde (1993), which uses the number of parasite individuals found in each host species and is based on a rank order of densities. A similar index can be defined using prevalence of infection rather than density. Numerical values for the indices, S_I, vary between 0 and 1, with values closer to 1 indicating a high degree of host specificity and closer to 0 a low degree. Poulin & Mouillot (2003) have also recently devised an index of host specificity. This index, S_{TD}, is phylogenetically based, and it relates to the phylo-genetic distance between pairs of host species used by a parasite species. It is independent of study effort, although the number of species used by a species of parasite correlates positively with the number of sampling records and ranges between a value of 1 when all hosts are of the same genus and 5 when all hosts belong to different classes. Using the same set of data on Canadian freshwater fishes, they showed that acanthocephalans used on average twice as many hosts as nematodes and three times as many as cestodes but the specificity indices did not differ significantly between groups (Table 4.3). They believed that the index was of value for comparative studies, and could be used to predict the likelihood of successful parasite invasions of a new habitat and host capture. However, like Rohde's index, no account was taken of the suitability status of each host species and all were treated as if they were preferred hosts. The index focused on the phylogenetic distance between hosts but took no account of the ecological distance, which of course is much more difficult to quantify. A later version of this index (Poulin & Mouillot, 2005b) attempted to combine the ecological approach of Rohde with their earlier phylogenetic approach. The revised index, $S_{TD}*$, measures the average taxonomic distinctness between host species and is weighted for prevalence in different host species. However, the new index still ignores the reproductive success of the parasite in each host species and the tacit assumption that all hosts are equally suitable for its reproduction remains.

   In reality, if one is comparing specificity between species or groups and is also considering the dangers of introductions, it is essential to incorporate the suitability of the host species in relation to the parasite's ability to reproduce into the evaluation of specificity. This is particularly the case with acanthocephalans in view of their employment of paratenic hosts, and their ability to survive but not reproduce in many species of hosts. The extent to which indices can give a distorted view of specificity is shown in Table 4.4. The acanthocephalan *Pomphorhynchus laevis* has been recorded from 16 species in the River Avon over a period of many years and in a large number of samples. Information on the density of the parasite and its growth in each species of fish is available, and so also is information on reproduction in each host. Just looking at the species list, i.e. records, suggests a very low degree of specificity. Inspection might also suggest that *Thymallus thymallus* is the most important host since parasite densities are on average greater in this host. Calculation of Rohde's index gives a value for S_I of 0.26, which suggests a fairly low degree of specificity: the value for the Poulin & Mouillot index S_{TD} is 3.84, which also suggests an intermediate to low degree of specificity. The use of reproductive criteria, however, indicates that only *Leuciscus cephalus*, *Barbus barbus* and *Oncorhynchus mykiss* can be regarded as preferred hosts, with *Salmo trutta* as a suitable host. The parasite in fact is found in every species of fish in the river, because all feed to a greater

Table 4.4. *Specificity of* Pomphorhynchus laevis *in the River Avon*

Host species	N	Prevalence	Maximum intensity	% females gravid
Barbus barbus	40	100.0	445	71.0
Leuciscus cephalus	8	100.0	359	67.2
Leuciscus leuciscus	30	83.3	45	4.2
Salmo trutta	15	86.7	141	14.7
Salmo salar	11	91.6	9	10.0
Oncorhynchus mykiss	15	93.3	52	43.4
Cottus gobio	154	100.0	10	0.5
Noemacheilus barbatulus	111	60.0	2	2.0
Thymallus thymallus	74	100.0	429	0
Rutilus rutilus	22	64.6	5	0
Gobio gobio	9	56.0	2	0
Anguila anguilla	12	16.0	3	0

Species where N < 8 omitted. N, number examined.
Source: Data from Hine & Kennedy (1974a) and Kennedy (2003).

or lesser extent upon the intermediate host *Gammarus pulex*, which is a key species in the food web. However, most of the host species belong to different genera and the extent to which they feed on the amphipod is very variable and accounts for the large differences in parasite density.

Another example of the necessity of obtaining information on reproductive success of a parasite in each species of host comes from the study by Holmes *et al.* (1977) and Leong & Holmes (1981) on Cold Lake in Alberta, Canada. Here again, every one of the 10 species of fish was infected with the acanthocephalan species *Metechinorhynchus salmonis*, which used only a single species of intermediate host *Pontoporeia affinis*. Infection levels in the fish species again reflected the extent to which they fed on this key species in the invertebrate community in the lake. The authors, however, were able to make estimates of the transfer of *M. salmonis* with ingested *P. affinis* to each species of fish and of the transfer of eggs to the crustacean from each species of fish. It is clear from these data that in fact only the whitefish, and especially *Coregonus clupeaformis* (Table 4.5) are the preferred hosts. Ashley & Nickol (1989) have similarly been able to show that in one locality *Leptorhynchoides thecatus* infects four species of definitive hosts: three species of *Lepomis* and one of *Micropterus*. However, *L. cyanellus* clearly emerges as the preferred host since it has the highest prevalence

Table 4.5. *Simplified scheme of derived relative flow rates for*
Metechinorhynchus salmonis *in a community of fishes in Cold Lake, Alberta*

Fish group	Flow of eggs from fish group to *Pontoporeia affinis*	Flow of cystacanths from *Pontorporeia affinis* to fish
Whitefish	0.518	0.702
Cisco	0.37	0.11
Lake trout	0.084	0.051
Coho	0.018	0.005
Pike, burbot, suckers, walleye, stickleback	0.011	0.131

Source: Data from Holmes *et al.* (1977) and Leong & Holmes (1981).

and intensity of infection and the highest proportion of gravid females: moreover, 46% of the flow from the intermediate host *Hyalella azteca* is through this species, and 54% of the egg flow comes from this species. Of the other species, *M. salmoides* and *L. gibbosus* can be considered suitable hosts on the basis of parasite flow to them and egg flow from them (between 21% and 29% in each case), whereas *L. macrochirus* is not a suitable host. Sometimes there may be several preferred species. Rauque *et al.* (2003) investigated the specificity of *Acanthocephalus tumescens* in a lake in Argentina. Using criteria of prevalence, abundance and number of gravid females they could recognise a continuum of six species, with *Diplomystes viedmensis* lying second and *Percichthys trucha* lying fifth. However, when parasite flow rates and egg output were calculated, *D. viedmensis* emerged as the least suitable host and *P. trucha* as the preferred host, with this and three other species being the main contributors to the acanthocephalan population (Fig. 4.3).

Recognition of criteria for specificity to intermediate and paratenic hosts poses particular problems. In the case of intermediate hosts, the great majority of data are derived from field observations, and these are essentially reports of the presence of cystacanths of a particular acanthocephalan species in a particular intermediate host. There is a tacit assumption that if a parasite develops to the cystacanth stage, it will be infective to the definitive host and therefore the arthropod in which it occurs is a suitable host. Since acanthocephalans either do or do not develop in a particular arthropod species, it is normally impossible to distinguish preferred from suitable hosts. In theory the status of an arthropod as intermediate host can be

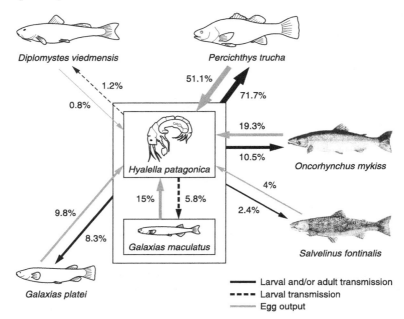

Fig. 4.3 Relative flow rates of *Acanthocephalus tumescens* in its hosts in Lake Moreno, Argentina. The thickness of the arrows indicates the relative values of parasite transmission and egg output. (From Rauque *et al.*, 2003, with permission.)

confirmed by laboratory experiments, but in practice this is unlikely since laboratory infections may at worst be impossible to achieve or at best uncontrollable. Attempts to infect *Gammarus pulex*, the intermediate host of *Pomphorhynchus laevis,* in the laboratory have almost invariably failed. Similar considerations apply to paratenic hosts. The ability of many acanthocephalan species to survive in an extra-intestinal site in unsuitable definitive hosts means that it is in practice very difficult to determine whether a particular host species is an unsuitable or accidental definitive host or a paratenic host. Even successful experimental infections are unlikely to give a clear result as they would fail to distinguish between status as a paratenic host and post-cyclic transmission from a definitive host. In practice laboratory infections have seldom been attempted and the presence of a cystacanth or undeveloped juvenile of a species known to infect raptors in a parenteral site in, for example, an amphibian is taken as evidence that the amphibian is a paratenic host. This may be true, but it does require experimental verification that the parasite can establish in the preferred definitive host.

Inspection of Table 2.3 does suggest that many species of acanthocephalans are more specific to their intermediate host than to their definitive host, e.g. *Mediorhynchus centurorum* and *Metechinorhynchus salmonis*, although some species appear to be quite widely specific to their intermediate host, for example *Macracanthorhynchus hirudinaceus* and *Polymorphus minutus*. However, such appearances may be deceptive. Several species of cockroach have been reported as intermediate hosts of *Moniliformis moniliformis*. Lackie (1975) had demonstrated that *M. moniliformis* could evade the cellular defence of *Periplaneta americana* and so this species could serve as a suitable intermediate host. Development was rare in other species of cockroach and she concluded that tolerant hosts were not closely related although they were all Blattoidea, whereas blaberids were unsuitable. Later experiments by Moore & Crompton (1993) and Moore & Gotelli (1996) confirmed this. A number of species of cockroach were exposed to eggs and to hatched acanthors by oral infection. The mode of infection had no influence on susceptibility. It was found that 12 species from six families and eight diferent genera were susceptible to infection (Fig. 4.4) but there was considerable interspecific variability in susceptibility in these species ranging from 13% to 100%. Other experimental infections have tended to confirm that specificity may be narrow. Nickol and Dappen (1982) confirmed that only young individuals of *Armadillidium vulgare* could be infected by *Plagiorhynchus cylindraceus*, and Uglem (1972) reported that eggs of *Neoechinorhynchus cristatus* could be ingested and would hatch in four species of ostracods but would only develop to the cystacanth stage in one. Similarly, attempts to infect other species of *Asellus* with eggs of *Acanthocephalus anguillae* and *A. lucii* have always failed (Brattey, 1986; A. R. Lyndon, unpublished) and both species appear narrowly specific to *A. aquaticus*.

4.3 SPATIAL VARIATIONS IN SPECIFICITY

The ability of so many species of acanthocephalan to utilise more than one species of preferred and suitable definitive or intermediate host allows flexibility in host usage in different localities. Thus, in the absence of a preferred species from a locality the parasite may be able to use an alternative or suitable species to survive and the flow of parasites through the host community will differ. In Cold Lake, for example, Leong & Holmes (1981) had identified whitefish, and especially *Coregonus clupeaformis*, as the key hosts to which most cystacanths of *Metechinorhynchus salmonis* flowed, in which most gravid females were

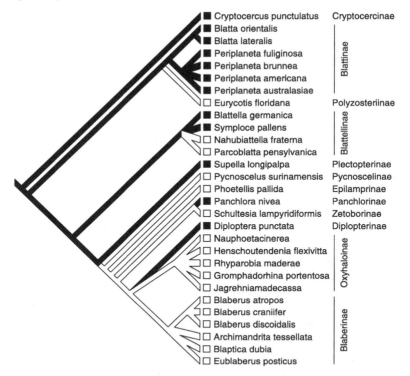

Fig. 4.4 Dendrogram of resistant and susceptible cockroach host species to experimental infections with *Moniliformis moniliformis*. Species arranged by subfamilies as indicated. Black, susceptible; white, resistant. (From Moore & Gotelli, 1996, with permission.)

found and from which most eggs were produced. However, in Lake Michigan Amin & Burrows (1977) found that *Osmerus mordax* and *C. hoyi*, both absent from Cold Lake, were together responsible for 38% of the gravid females; but the greatest single proportion of gravid females, 30.2%, was found in *Oncorhynchus kisutch*, a species present in Cold Lake but there containing only 1.8% of the gravid females (Table 4.6). The differences in host usage between the two localities thus reflected not only the presence of different fish species in the communities but also differing degrees of predation upon *Pontoporeia affinis*, the only intermediate host in both localities.

Different patterns of host usage in different localities can also be seen in *Pomphorhynchus laevis*. Its preferred definitive hosts in fresh water across its range are *Barbus barbus* and *Leuciscus cephalus*, but in the absence of either one of these two species it can utilise the other.

Table 4.6. *Host utilisation by* Metechinorhynchus salmonis *in two American lakes*

Host	Cold Lake			Lake Michigan		
	N	A	% G	N	A	% G
Clupeidae						
Alosa pseudoharenganus	0	0	0	149	0.2	0.1
Salmonidae						
Coregonus artedii	757	2.8	37.0	0	0	0
C. clupeaformis	836	169.2	51.8	8	22.0	1.8
C. hoyi	0	0	0	79	12.5	16.4
Oncorhynchus kisutch	291	189.5	1.8	15	140.3	30.2
O. tshawytscha	0	0	0	6	53.3	6.2
O. mykiss	0	0	0	8	38.1	4.3
Salmo trutta	0	0	0	16	16.4	3.8
Salvelinus namaycush	35	421.0	8.4	7	81.9	10.9
Osmeridae						
Osmerus mordax	0	0	0	446	3.1	21.7
Gadidae						
Lota lota	29	174.0	1.0	1	18.0	0.3
Cottidae						
Cottus cognatus	0	0	0	37	6.8	4.1

N, number of hosts examined; A, parasite abundance; %G, percentage of gravid females.
Species which contained no gravid females and those in which the % of females gravid < 1.0 are omitted.
Source: Data on Cold Lake from Leong & Holmes (1981) and on Lake Michigan from Amin & Burrows (1977).

In some fresh water localities it uses *Salmo trutta*, whereas in estuarine localities, or waters of raised salinity, it uses *Platichthys flesus* (Table 4.7). It appears also to use the local species of *Gammarus* as its intermediate host. This ability to use different hosts in different localities is equally evident on a much smaller scale, in southern England for example (Table 4.8). In the River Culm, from which barbel are absent, it uses *L. cephalus* as its preferred host and *Leuciscus leuciscus* as an additional suitable host, but in the River Avon where both species of preferred host are present, *L. leuciscus* is a minor and unimportant host. In the

Table 4.7. *Hosts of Pomphorhynchus laevis in Western Europe*

Country	Intermediate host	Definitive host	Distribution
Ireland[a]	*Gammarus duebeni*	*Salmo trutta*	Widespread
Scotland[b]	*Gammarus duebeni*	*Salmo trutta*	Localised
England[c]	*Gammarus pulex*	*Barbus barbus*	Very local
		Leuciscus cephalus	
France[d]	*Gammarus fossarum*	*Barbus barbus*	Localised
Italy[e]	*Echinogammarus stammeri*	*Leuciscus cephalus*	Localised
		Salmo trutta	
		Barbus tyberinus	
Austria/Hungary[f]	*Gammarus roeseli*	*Barbus barbus*	R. Danube
Czech Republic[g]			
Slovak Republic[g]	*Gammarus balcanicus*	*Leuciscus cephalus*	Uncertain
Baltic & North Seas: estuaries[h]	*Gammarus locusta* *Gammarus zaddachi*	*Platichthys flesus*	Widespread

Sources: Data from:
[a]Fitzgerald & Mulcahy (1983);
[b]Kennedy unpublished,
[c]Kennedy *et al.* (1989);
[d]Van Maren (1979a,b);
[e]Dezfuli *et al.* (1991a, 1999);
[f]Moravec *et al.* (1997);
[g]Kral'ova-Hromadova *et al.* (2003);
[h]Kennedy (1984a).

absence of both preferred hosts in the River Otter, it uses *Salmo trutta* as preferred host and *Cottus gobio* as an additional host, yet this species is a minor, insignificant host in the other two localities. This flexibility in host usage enables the parasite to take the maximal advantage of any locality in which it is found. Other species of *Pomphorhynchus* may be less flexible. In Patagonia, *P. patagonicus* utilises only *Hyalella patagonica* as an intermediate host and the endemic fish species *Patagonina hatcheri*, *Percichthys trucha* and *Galaxias platei* as definitive hosts, with either of the first two species being preferred in different localities (Semenas *et al.*, 1992; Trejo, 1992, 1994; Ubeda *et al.*, 1994). However, the introduced *Oncorhynchus mykiss* is a completely unsuitable host, whereas it is a very suitable one for *P. laevis*. The North American species *P. bulbocolli* similarly shows narrow specificity to its intermediate host *Hyalella azteca* (Amin, 1978, 1987), but wide specificity to its preferred

Table 4.8. *Maturation of* Pomphorhynchus laevis *in three English rivers*

Locality and hosts	N	Prevalence (%)	Abundance Mean ($\pm$SD)	Gravid females	
				%	Mean wt (g)
R. Avon					
Barbus barbus	18	100.0	169.0 (105)	71.0	64.4
Leuciscus cephalus	8	100.0	68.0 (115)	67.2	62.2
Leuciscus leuciscus	30	83.3	5.3 (7.2)	4.2	14.9
Oncorhynchus mykiss	15	93.3	8.2 (18.4)	43.4	38.8
Salmo trutta	15	86.7	13.8 (27.1)	14.7	38.3
Cottus gobio	154	100.0	4.3 (6.1)	0.5	12.6
Noemacheilus barbatulus	111	60.0	1.1 (1.5)	2.0	9.2
R. Culm					
Leuciscus cephalus	15	100.0	74.1 (149.6)	61.0	47.0
Leuciscus leuciscus	7	87.5	1.4 (1.5)	30.0	35.7
R. Otter					
Salmo trutta	22	100.0	23.8 (16.9)	71.0	27.5
Cottus gobio	32	50.0	1.8 (2.4)	23.5	13.5
Platichthys flesus	16	43.6	1.6 (2.9)	4.6	15.3
Anguilla anguilla	22	27.2	2.1 (9.4)	2.1	40.1

Species in which the parasite occurs but does not become gravid are omitted.

definitive hosts, the identity of which changes with locality, and it is also able to use paratenic hosts (Amin, 1987).

Sometimes differences in host usage are not easy to understand. *Acanthocephalus clavula* can use several species of fish as definitive hosts in the British Isles, though it shows a preference for eels *Anguilla anguilla* (Chubb, 1964, 1967). Its only known intermediate host there is the isopod *Asellus meridianus*. The distribution of the parasite and infection levels in fish species relate closely to the distribution of the isopod and the extent to which fish feed on this species. In Italy, it normally uses the local species *Asellus coxalis* as intermediate host, but in the River Adige adults infected eels but the cystacanths were found in *Echinogammarus pungens* (Dezfuli *et al.*, 1991b). Schmidt (1985), in fact, lists seven species of gammarid from which it has been recorded in Europe. Adults of this species may be found in brown trout *Salmo trutta* and occasionally they may be abundant in this host (Chubb, 1967).

Brown trout can be a suitable host for *A. clavula* (Lyndon & Kennedy, 2001; Byrne *et al.*, 2004) though this happens very rarely. Unusually, in Clogher Lake in the west of Ireland, 86% of brown trout were infected with a mean of 53 parasites per fish, even though only 2% of females attained full reproductive maturity (Byrne *et al.*, 2004). Trout here were clearly an accidental host, as 97% of eels were infected and 61% of females became fully mature. The unusually high levels in trout were considered to reflect their feeding preference for *A. meridianus* and the absence of an acanthocephalan competitor, *P. laevis*, from the lake.

Other species such as *Leptorhynchoides thecatus* can also show great spatial flexibility in host usage (Ashley & Nickol, 1989). It uses *Hyalella azteca* as its intermediate host, but in one reservoir studied the green sunfish *Lepomis cyanellus* was identified as the principal definitive host based on the flow of parasites, infection levels and proportion of gravid females. In other localities smallmouth bass *Micropterus dolomieui* is often regarded as the principal definitive host. In the Great Plains region, where this species is less common than further east, the largemouth bass *M. salmoides* is generally the preferred host. In the reservoir investigated, there were no smallmouth bass but *M. salmoides* were common. Nevertheless, although some parasites did mature in largemouth bass, the green sunfish was clearly the preferred host and *L. thecatus* was rare in bluegill sunfish in which it is common elsewhere. In a later study, Olson & Nickol (1996) showed that both green sunfish and largemouth bass were equally good hosts in laboratory infections, so the difference in usage in the reservoir must relate to local ecology. The authors commented that single fish species seldom assumed identical roles in the usage by acanthocephalans in different localities.

On an even more local scale, Jackson & Nickol (1981) investigated the ecology of the specificity of *Mediorhynchus centurorum* in woodpeckers. In the locality studied, the acanthocephalan was found only in the red-bellied woodpecker *Melanerpes carolinus*, although seven other species of woodpecker were examined, including four from a site where 62% of *M. carolinus* were infected. In this site the intermediate host was the woodroach *Parcoblatta pennsylvanica*; when these four species were fed infected roaches in the laboratory, they developed patent infections of *M. carolinus*.

The explanation for this distribution appeared to lie with the roaches. No roaches from swampy areas of the locality were infected with the acanthocephalan, but 9% from the pasture area were; all nests of *M. carolinus* were found in the pasture area while those of the four other species were in the swampy area. Moreover, nests of *M. carolinus* were not particularly sanitary, eggs of *Mediorhynchus centurorum* were concentrated around the nests and the accumulation of food and faeces was attractive to roaches. Woodroaches were also the main food item offered to nestlings of the red-bellied woodpecker but were not offered to Red-headed nestlings. Thus, small differences in nesting behaviour and diet, i.e. in ecology between closely related host species, were responsible for major differences in specificity.

4.4 STRAIN SPECIFICITY

Variation in specificity is clearly adaptive to an acanthocephalan species in enabling it to survive in localities where some of its preferred hosts are absent. However, the flexibility in host usage may be more apparent than real and in some cases acanthocephalans may be more specific than is generally believed. There is evidence that acanthocephalans may have adapted to local hosts to such a degree that they have come to constitute a separate strain. These strains may not be distinguishable on the basis of morphological features, but only become apparent in the light of differences in host usage, position in host and other biological and distributional characteristics. It may also require molecular methodology to confirm the distinctions.

In their studies on *Polymorphus minutus* in ducks, Hynes & Nicholas (1958) had observed that the three native species of *Gammarus* could be found infected with cystacanths of *P. minutus*, whereas the introduced *G. tigrinus* was resistant to infection. However, they were also able to demonstrate experimentally that when eggs were fed to *Gammarus* of the same species from which cystacanths were obtained to infect ducks, development proceeded normally whereas when they were fed to different species of *Gammarus* partial or complete failure resulted (Table 4.9). This was clear evidence of a much greater degree of specificity than had been initially apparent, and they concluded that it provided evidence for the existence of distinct strains of *P. minutus*.

Aho *et al.* (1992) investigated the status of congeneric species of *Neoechinorhynchus* in the turtle *Trachemys scripta*. It was common to find two to four congeners in a single turtle, of which three species exhibited a high degree of overlap in their position in the intestines. The two most likely explanations for this were that there were subtle differences in resource requirements between the species, or that they actually constituted a single, but variable, species. Using electrophoretic methodology on 24 enzymes, a genetic analysis of six loci excluded the single species hypothesis but was consistent with the existence of

Table 4.9. *Experimental infections of* Gammarus *species with different strains of* Polymorphus minutus

Species of origin of *P. minutus*	Species tested	% normally infected	% with black spots	% with acanthors and spots	% not infected
Gammarus pulex	*G. pulex*	87.0	0	0	13.0
	G. duebeni	23.0	67.0	12.0	22.0
	G. lacustris	57.0	8.0	3.0	39.0
Gammarus duebeni	*G. pulex*	0	48.0	0	52.0
	G. duebeni	93.0	4.0	1.0	5.0
	G. lacustris	59.0	44.0	18.0	15.0
Gammarus lacustris	*G. pulex*	25.0	57.0	14.0	32.0
	G. duebeni	0	100.0	0	0
	G. lacustris	98.0	23.0	21.0	0

Source: Reproduced from Hynes & Nicholas (1958) with permission of the Liverpool School of Tropical Medicine.

four congeneric species. It was also clear that any individual turtle could harbour individuals of a given species that were genetically different. Attempts to designate these species on morphological criteria were not error free, and on the whole taxa defined on electrophoretic characters were not congruent with those defined on morphological ones. Genetic variation on this scale could cause confusion in ecological studies as subtle differences in the composition and distribution of species have the potential to confound community analysis. They certainly make it even more difficult to discern patterns in specificity. A similar situation was identified by Buron *et al.* (1986). Based on an electrophoretic study of 18 enzymes, their work described a new species, *Acanthocephalus geneticus*, from *Arnoglossus laterna* from the Mediterranean littoral. It was morphologically very similar to *A. propinquus* from *Gobius niger* and indeed it had previously been considered that the slight morphological differences were part of the normal variation in *A. propinquus* which was believed to be fairly non-specific and capable of infecting *A. laterna*. However, it was now clear that there were two species, which differed considerably in their specificity.

Morphological variability has also caused problems in the genus *Echinorhynchus*. Shostak *et al.* (1986) studied the variability in three species from fish in Canada, where the wide host range and mixed species infections made identifications very difficult in the absence of data on morphological variability from geographically distinct populations. All characters showed great variability and high levels of interspecific overlap. Morphological variability between different populations in different lakes was greater than could be accounted for by host species and environmental differences. They therefore suggested that the geographic variability could be a consequence of restricted gene flow between isolated populations causing geographic isolation of morphological variants, i.e. strain formation. In Europe, Vainola et al. (1994) conducted an allozyme study of some European members of the genus *Echinorhynchus*. This revealed high levels of inter specific and intraspecific genetic divergence as well as the existence of new biological species. There was a clear distinction between the marine *E. gadi* and the freshwater *E. salmonis* but this did not agree completely with the current views on the concepts of *Echinorhynchus* and *Metechinorhynchus* as separate genera. Samples of *E. gadi* from northern waters included three distinct, partially sympatric biological species, and *E. bothniensis* was considered to be a complex of three freshwater taxa, which all used the relict species of *Mysis* as intermediate hosts. A freshwater population was related to *E. bothnensis*, but was probably specifically distinct from it. There was also evidence of differences between groups reflecting post-glacial events. Thus, what was originally thought to be three species each showing wide specificity is now apparently a complex of at least seven taxa with narrower specificity.

Post-glacial events have also been invoked to explain the distribution of *Pomphorhynchus laevis* in the British Isles (Fig. 3.3 and Table 3.10). On biological grounds, including distribution, host usage and position in host, Kennedy et al. (1989) hypothesised that there were three distinct strains of *P. laevis* in the British Isles. The marine strain originated post-glacially in the Baltic and North Sea, where it infected flounders *Platichthys flesus*, with a preference for the rectum (Kennedy, 1984a) and *Gammarus locusta* as its intermediate host. It can also be found in the tidal Thames, where it uses *G. zaddachi* or *G. salinus* as an intermediate host (Munro et al., 1989; El-Darsh & Whitfield, 1999; Guillen-Hernandez & Whitfield, 2001, 2004) and overlaps in distribution with the freshwater strain. Using molecular methodology on genomic DNA, Munro et al. (1990) found the marine strain to be very similar to the English freshwater strain but with sufficient differences to justify considering it to be a separate strain. The English strain was considered to have been separated from the Rhine and to have originally been restricted to the Thames, having been spread from there

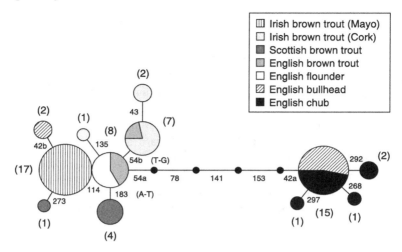

Fig. 4.5 Network of *Pomphorhynchus laevis*, cytochrome c oxidase (subunit I) from Ireland and Britain. Partial sequences of 295 base pairs in length were aligned. Each line (delineated by black dots) represents one base difference between samples. Large numbers in parentheses represent sample sizes. Smaller numbers show the position of each mutation, with transversions in parentheses. (From O'Mahony *et al.*, 2004b.)

Table 4.10. *Percentage nucleotide identity or ITS-1/ITS-2 in five samples of* Pomphorhynchus laevis *and* P. lucyi

Acanthocephalans	Slovakia	Italy + Czech	Italy
	Chub + minnow	Chub + chub	Barbel
Chub Italy + Czech	88.7/91.3		
Barbel Italy	87.3/91.3	98.4/100.0	
P. lucyi	56.1/65.3	52.4/63.1	51.8/63.1

Source: Modified from Kral'ova-Hromadova *et al.*, (2003) with permission.

to other rivers by stocking with barbel, *Barbus barbus*. It uses *G. pulex* as intermediate host.

The Irish strain, by contrast, uses *Salmo trutta* as definitive host and *G. duebeni* as intermediate host. It has been suggested that it was introduced to Ireland through fish stocking, and in the absence there of chub or barbel used its ability to reproduce in trout to survive (Kennedy *et al.*, 1989). In addition to host usage, there are morphological features that can distinguish the Irish and English parasites (O'Mahony *et al.*, 2004a). However, molecular tests of this strain hypothesis by sequencing a mitochondrial gene and using parasites from England, Ireland and Scotland has revealed that all parasites from English cyprinid fish are very similar. Parasites from Irish and Scottish trout are similar to each other but distinct from the English material (O'Mahony *et al.*, 2004b) (Fig. 4.5) thus confirming the existence of two freshwater strains in Britain and Ireland but suggesting the difference may be more host than geographically based. It is also important to realise that the strains still show wide host specificity: they may exhibit different host preferences but they still retain the ability to infect a range of fish species. Only the marine/Baltic strain appears to show narrow specificity.

While the taxonomy, biology and distribution of the strains in the British Isles have been studied in some detail, the possibility of strain formation in continental Europe has until recently been largely overlooked. However, comparison of *P. laevis* material from the Czech and Slovak Republics using isoenzyme and morphological analyses suggested the possibility of strain differences on the continent itself (Dudinak & Snabel, 2001). A subsequent study of *P. laevis* from central Europe was conducted by Kral'ova-Hromadova *et al.* (2003) using the internal transcribed spacers (ITS-1 and ITS-2) of the ribosomal RNA gene. The nucleotide sequences of ITS regions of *P. laevis* from two different hosts, minnow *Phoxinus phoxinus* and chub *Leuciscus cephalus*, and two different and distant localities, a river and a lake, in the Slovak Republic were found to be 100% identical. Samples from chub from the Czech Republic and Italy were also mutually identical. However, the sequences from the Slovak and Czech parasites were significantly different (Table 4.10) with high levels of variation. A sample from *Barbus tyberinus* from Italy was very similar to the sympatric isolate from chub (Table 4.10), with the small difference probably representing intra-population variation. Comparison of sequences with the North American species *P. lucyi* showed very distinct and significant differences (Table 4.10), confirming that the two species are indeed distinct. From these findings, the authors believe that there are at least two strains of *P. laevis* on the continent which are well defined on a molecular basis but which differ only in minor details of morphology. The relationship between these strains and host specificity is far less clear than in the British Isles, as both strains can use chub as a preferred definitive host. Although the Slovak strain uses *Gammarus balcanicus* as intermediate host in both localities, the Czech/Italian strain uses *Echinogammarus stammeri* in Italy and *Gammarus roeseli* in the Czech Republic. The full extent of the distribution of neither strain is yet known, and further studies are clearly needed, but it seems likely that other strains will come to be recognised on the continent.

Overall, it is not easy to detect patterns in specificity throughout the Acanthocephala when species distinctions are based on morphological criteria. Some species show narrow specificity to their hosts and some broad, but there is little or no correlation between the degrees of specificity shown to their two or three hosts. Specificity does often seem to be narrower to the intermediate than to the definitive host, but this is not consistent. The overall impression is of flexibility. Some acanthocephalans have escaped the rigidity of their two-host life cycle by adding hosts in series, either by incorporating a paratenic host into the sequence or by being able to undergo post-cyclic transmission, but most seem in addition, or instead, to have added hosts in parallel at each stage to increase their versatility. Only rarely is there any phylogenetic relationship between these hosts, and phylogenetic specificity, so common in cestodes and monogeneans, is relatively uncommon in acanthocephalans. The flexibility this gives them enables them to survive in locations from which their preferred or even most suitable hosts are absent. It can be viewed as an insurance policy against habitat and host changes in time and space and as a means of minimising the risks of local extinction.

Just how much potential flexibility there may yet be in the genomes that is so far unexpressed remains to be determined. At the same time, studies on strains and molecular techniques have suggested that specificity may be narrower than first thought, and that local speciation and subspeciation may be far more widespread than has hitherto been realised. It is thus possible that many apparently widely specific species may yet turn out to be a complex of more narrowly specific sibling species. However, whichever way you look at it, the acanthocephalans show enormous variation in specificity and their strategy contributes in no small way to their distribution and abundance and success.

5

Host—parasite interactions

5.1 DISTRIBUTION OF ACANTHOCEPHALANS WITHIN HOSTS

5.1.1 Distribution in individual hosts

Although adult acanthocephalans are found in the alimentary tract of their vertebrate definitive hosts, they do, in common with other groups of intestinal parasites, exhibit preferences for particular regions of that tract. Crompton (1973) has reviewed the sites occupied by parasitic helminths in the alimentary tract of vertebrates and concluded that the majority of species show a clear preference for a particular region of the alimentary canal even if their distribution extends both anterior and posterior to this region. The preferred site may relate to the physico-chemical conditions in the alimentary tract (Crompton, 1970, 1973; Taraschewski, 2000), to particular stimuli required for eversion of the larval stage (Kennedy et al., 1976), to specific nutritional requirements (of which we know little or nothing) or to interspecific interactions (Holmes, 1973). The site may extend in response to increasing parasite density, i.e. crowding (Kennedy & Lord, 1982: Kennedy, 1984a), or decrease in response to competition from other species (Holmes, 1973). It may also differ in different host species (Kennedy, 1985b), or even strains. The marine strain of *Pomphorhynchus laevis*, for example, shows a clear preference (Kennedy, 1984a) for the rectum in its preferred host, the flounder *Platichthys flesus*, whereas both English and Irish strains prefer the posterior intestine in their principal hosts and in flounders (Fitzgerald & Mulcahy, 1983; Kennedy et al., 1976, 1978). Furthermore, the English and Irish strains show small differences in their preferred site in trout *Salmo trutta* (Molloy et al., 1993, 1995). Site preference is exhibited by all species of acanthocephalans in all classes of vertebrate definitive host, and indeed the preferred site corresponds exactly to the 'zone of viability' of *Moniliformis moniliformis* in rats as described by Burlingame and Chandler (1941). The relationship between this preferred site and the niche of a species is discussed in more detail in Chapter 7, but at this stage it is sufficient to recognise that all species do exhibit a preference and that the majority of an infrapopulation (all the individuals of a species in a single host individual) is likely to be found in this site.

Even in preferred definitive hosts, it is not uncommon in some species to find individuals in the body cavity. Adults of some species, such as *Pomphorhynchus* species, penetrate deeply into the intestinal wall, and some individuals penetrate it completely, pass through it and come to lie in the body cavity (Hine & Kennedy 1974a). These do not develop any further but eventually become encapsulated and lysed by host reactions (Taraschewski, 2000). As far as is known, they are not able to develop following post-cyclic transmission. Extra-intestinal individuals are seldom found in species that exhibit only shallow penetration of the intestinal wall. Acanthocephalans in paratenic hosts also normally occupy extra-intestinal positions. They may be found loose in the body cavity or wholly or partially embedded in internal organs such as the liver (Taraschewski, 2000) or encapsulated in the body cavity as is common in species of *Corynosoma* (Valtonen, 1983) but they do not appear to show any more distinctive site preference than this.

Cystacanths in intermediate hosts are always found in the haemocoel of the arthropod. They may be enclosed in membranes, but as long as they are still viable they normally appear to be free in the cavity. Although they may exert pressure on some of the internal organs, particularly in multiple infections of small host individuals, it is only in large hosts such as crabs that cystacanths can appear to be attached to them. It is not possible to recognise a preferred site for them more precisely than this.

5.1.2   Distribution within a host population

In any host population, some individuals will be more and some less heavily infected than others. A uniform distribution, in which every host individual harbours exactly the same number of parasites, is very unusual amongst parasites. The normal situation encountered is that some individuals are very heavily infected, some harbour intermediate levels of infection and some are uninfected. Observed distributions are dynamic and are generated by opposing factors, some of which

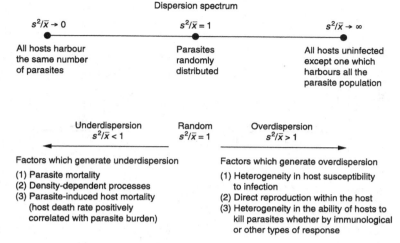

Fig. 5.1 Diagrammatic representation of a dispersion spectrum denoting the factors which create underdispersion and the factors which generate overdispersion. (From Anderson & Gordon, 1982.)

promote aggregation and some uniformity (Anderson & Gordon, 1982). The dispersion of parasite numbers within a host population can thus be viewed as a spectrum (Fig. 5.1). At one extreme all hosts harbour the same number of parasites and at the other all the parasites are located in a single host individual population. The terms used to describe these extremes in the parasitological literature are under-dispersion and overdispersion respectively (these are often used to have the reverse meanings in the literature on free-living animals, which can be a source of considerable confusion). At any moment in time, a variety of demographic and environmental factors act upon the parasite population and as these change so does the dispersion pattern. The degree of dispersion can be quantified, and two indices are commonly used to describe the dispersion. The first is the variance to mean ratio of parasite abundance in the population. This equals unity for a random dispersion, is less than unity for an underdis-persed population and exceeds unity for an aggregated (overdispersed) population (Fig. 5.1). The population can also be characterised by fitting the dispersion to a negative-binomial model and calculating the parameter k: the lower the value of k, the more aggregated the dispersion.

Under most conditions and for most of the time, acantho-cephalan populations are overdispersed: indeed, Crofton (1971a) has

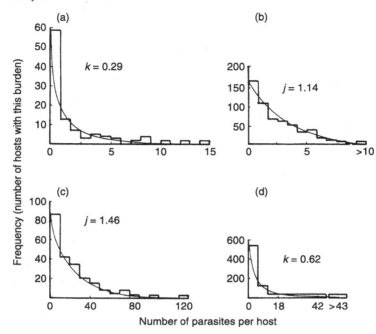

Fig. 5.2 Frequency distributions of the numbers of acanthocephalan parasites per host. (a) *Acanthocephalus clavula* in *Gasterosteus aculeatus* (Pennycuick, 1971); (b) *Polymorphus minutus* in *Gammarus pulex* (Crofton, 1971a); (c) *Moniliformis moniliformis* in experimentally infected *Periplaneta americana* (Holland, 1983); (d) *Pomphorhynchus laevis* in *Leuciscus leuciscus* (Kennedy & Rumpus, 1977). (From Dobson & Keymer, 1985.)

suggested that overdispersion should form part of the definition of a parasite. At very low densities (when close to extinction or when a species has just invaded a locality), the dispersion may be close to random but this is generally a temporary situation as the population level continues to change. Overdispersion of acanthocephalans is characteristic of both invertebrate and vertebrate hosts (Fig. 5.2). The degree of overdispersion may also vary in space.

Using data generated by Hynes & Nicholas (1963) on *Polymorphus minutus* levels in *Gammarus pulex* in a small stream above and below a point source of the acanthocephalans in a duck pen, Crofton (1971b) showed that overdispersion was highest close to the duck pen and declined downstream (Table 5.1). As infection levels change with host age, so also will the degree of dispersion. The degree of dispersion of parasites has an important bearing on the probability of a parasite being mated. Clearly, the greater the number of parasites in an

Table 5.1. *Changes in the degree of overdispersion of* Polymorphus minutus *in* Gammarus pulex *in a small stream above and below a duck pen*

Station	Location (m)	Mean abundance	k	$\chi2$
Above	50	0.38	0.33	sig.
Below 1	10	2.36	1.14	sig.
2	250	1.42	1.58	sig.
3	800	1.32	3.05	ns
4	2500	0.27	0.60	ns

k, parameter of negative binomial model; sig., significant; ns, not significant. *Source:* Modified from Crofton (1971a), in turn based on data from Hynes & Nicholas (1963).

individual host, the greater will be the probability of one meeting another of the opposite sex, and if, for example, all the parasites were in a single host individual, it is evident that the probability of mating would be considerably higher than if most hosts harboured only two or three parasites.

Infection levels will also change with host age. Age-intensity profiles may change throughout the lifespan of a host as demographic, physiological and ecological factors change. A saturated profile may result from the preferred site of an acanthocephalan species in the intestine being fully occupied and so setting a limit on parasite intensity. A convex profile may result from increased parasite mortality over time, whether in response to a host reaction or natural ageing, without further infection. A sigmoid profile may result from a decline in transmission rate, with or without parasite death, which in turn could result from a change in diet with age. Considerable attention has been focused upon convex curves, as a convex age-intensity curve concomitant with a decline in the degree of overdispersion in older age classes of hosts may, under some circumstances (Anderson & Gordon, 1982), suggest parasite-induced host mortality (e.g. Fig. 5.3). Examples of age-intensity profiles from natural populations of acanthocephalans are shown in Fig. 5.4 (the data are shown as host size rather than age, but in fish size and age are closely correlated). However, both Anderson & Gordon (1982) and Duerr et al. (2003) have emphasised the difficulties in interpreting age-intensity profiles. These can be affected by processes underlying the distribution of parasite numbers in host populations. These include: age-dependent exposure to parasites; parasite-induced host mortality; heterogeneity in host susceptibility; clumped infections

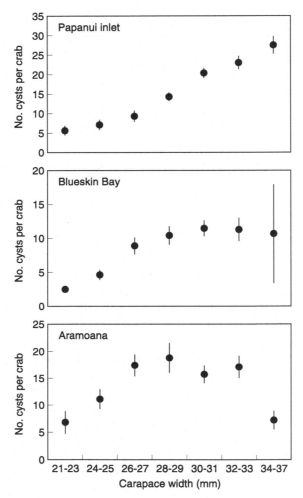

Fig. 5.3 Acanthocephalan, *Profilicollis* sp., induced host mortality in the mud crab *Macrophthalmus hirtipes*. Mean larval intensity (+/− S.E.) as a function of size classes is shown for three different intertidal populations from South Island, New Zealand. The levelling out and decrease in mean parasite load in the larger size classes in Blueskin Bay and at Aramoana respectively suggest parasite-induced host mortality among the larger and more heavily infected crabs. (From Mouritsen & Poulin, 2002.)

and density-dependent parasite mortality or establishment. Any or all of these processes can operate alone or together, to generate some very peculiar patterns that can in turn lead to ambiguous interpretations. Despite these cautions, there has as yet been no claim that age-intensity

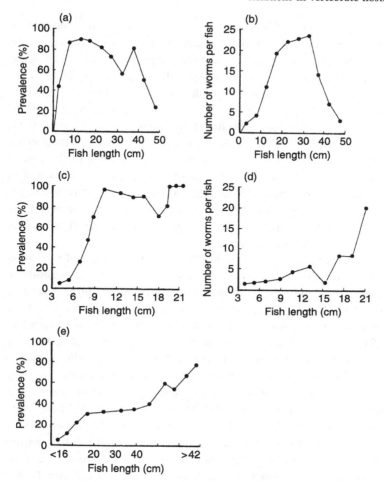

Fig. 5.4 The prevalence and intensity of acanthocephalan infections in relation to host size. (a) and (b) *Octospinifer macilentus* and *Pomphorhynchus bulbocolli* in whitesuckers (Muzzall, 1980); (c) and (d) *Neoechinorhynchus saginatus* in fallfish (Muzzall & Bullock, 1978); (e) *Neoechinorhynchus rutili* in sticklebacks (Walkey, 1967). (From Dobson & Keymer, 1985.)

curves have provided evidence of acanthocephalan-induced vertebrate host mortality.

5.2 RELATIONS IN VERTEBRATE HOSTS

Adult acanthocephalans attach to the intestines of their definitive hosts by embedding their probosces in the intestinal wall. These probosces with their rows of hooks appear capable of inflicting considerable damage to the gut wall and so upon their hosts. The significance of the differences in patterns of hook arrangement and numbers of hooks are unknown: all the species illustrated in Fig. 1.1 can be found in the same species of host. Although each has a preferred host, both *Acanthocephalus anguillae* and *Pomphorhynchus laevis* use *Leuciscus cephalus* as a preferred host and can be found in similar regions of the intestine, yet they exhibit completely different patterns of hook arrangement. It would seem likely that damage would be enhanced in the case of those species that possess a bulb (e.g. species of *Pomphorhynchus*) and by the fact that acanthocephalans aggregate in a preferred region of the gut. Moreover, infection intensities can in some cases be very high: infrapopulation densities of some species may range from several hundred to a thousand.

The effects of acanthocephalans on their hosts and vice versa can be considered as continuous variables between the extremes of susceptibility and total resistance to infection. There is little doubt that acanthocephalans do cause local damage to their host intestines (Bullock, 1963; Taraschewski, 2000). This is frequently manifested as inflammation in the region of infection, and is often accompanied by the production of fibrous tissue around the proboscis and/or proboscis bulb. When infections are heavy, whole areas of the intestine may appear damaged (Hine & Kennedy, 1974a) as absorbtive tissue is re-placed by fibrous tissue. Acanthocephalans can clearly have more general effects on their hosts, in that they can provoke host immune responses. Harris (1972) has shown that adult, fully mature *Pomphorhynchus laevis* does provoke immune antibody responses in its preferred fish hosts, even though the parasites appear to be unaffected by the antibodies. In rats, by contrast, *Moniliformis dubius* provokes an immune response that can reduce establishment rates of parasites at high densities (Andreassen, 1975a,b). It would also seem likely that individual acanthocephalans penetrating the gut wall completely and migrating though it into the body cavity could carry with them bacteria or other matter capable of causing pathogenic changes in the cavity. Nevertheless, these hosts so heavily infected in the intestines and with parasites in the body cavity are frequently found to be in excellent condition and in apparent good health; Hine & Kennedy (1974a) could detect no difference in the condition factors between uninfected fish and fish carrying infections of *Pomphorhynchus laevis*. Similarly, Szalai & Dick (1987) were able to show that *Neoechinorhynchus carpioidi* could cause nodule formation and local damage to the intestine of its fish host, but were unable to find any more general effects on fish.

There are reports of damage to host individuals caused by acanthocephalans. Bullock (1963) was clearly of the opinion that *Acanthocephalus jacksoni* was pathogenic to several species of trout as evidenced by the histopathology changes in infected fish and·by their condition. Bertocchi & Francalanci (1963) also reported serious damage to trout in a breeding farm resulting from *Echinorhynchus truttae* infections. Mortality was particularly high in young trout, but fish that survived showed some loss of weight. It does appear that in this case infection levels may have been unusually high under the farm conditions. In other cases, parasite-induced host mortality has been deduced from the finding of dead hosts with heavy infections, for example dead eider ducks with high burdens of *Profilicollis botulus*. Liat & Pike (1980) and Thompson (1985c) considered that epizootics could occur and might be due to eiders selectively eating large, infected crabs when numbers of alternative food sources such as mussels crashed. Nicholas & Hynes (1958) also considered that domestic ducks could suffer mortality from infections with *Polymorphus minutus* following exposure to large numbers of cystacanths. Nickol (1985) reviewed many examples of host death being attributed to acanthocephalan infections, and concluded that the pathological significance of acanthocephalans in dead or dying animals is difficult to interpret as densities are often no higher than those in surviving individuals which show no deleterious effects. By contrast, he does consider that in some cases acanthocephalans may cause epizootics when infection levels increase suddenly for whatever reason. In addition to the above examples, he cites Schmidt *et al.* (1974) who reported a major decline in a population of *Cottus bairdi* apparently associated with unusually heavy levels of *Acanthocephalaus dirus* infections and he also reported possible effects of *Polymorphus* species on young swans. The report of Mazzi & Bakker (2003) that *Pomphorhynchus laevis* imposes a survival cost on three-spined sticklebacks *Gasterosteus aculeatus* proportional to the severity of infection needs to be seen in perspective. Sticklebacks are neither a normal nor a preferred host for this parasite, which is unlikely to be adapted to it. Only some sticklebacks were susceptible to infections and the relative density and biomass of parasites in relation to the size of the fish is far higher than that likely to be encountered in natural situations of *P. laevis* in chub or barbel.

Overall, therefore, it appears that acanthocephalans may on occasion reach densities at which they can cause epizootics, but these occasions would appear to be the exceptions. The general situation is that while they do cause local damage to their definitive vertebrate hosts, they are not normally or regularly responsible for vertebrate host mortalities. Acanthocephalans do not normally appear to alter the behaviour of their vertebrate hosts in any way as a response to or defence against infection (Hart, 1994), and there is only a single report of an acanthocephalan altering the behaviour of a vertebrate, in this case the swimming and diving behaviour of a skink (Daniels, 1985).

There is also no evidence that acanthocephalans in any way alter the behaviour of their paratenic hosts to facilitate infection.

5.3 RELATIONS IN INVERTEBRATE HOSTS

Because the size of a cystacanth relative to that of many arthropods is large, it might be expected that acanthocephalans may have a pathogenic effect upon their invertebrate host. This is indeed often the case, especially when the intermediate host is an isopod or amphipod. A fully developed cystacanth in the haemocoel is surrounded by an envelope or capsule. In the case of *Moniliformis moniliformis*, the envelope helps the parasite avoid recognition by its *Periplaneta* host (Loker, 1994). A healthy cystacanth of *Polymorphus minutus* is similarly protected from recognition in its *Gammarus* host and it does not provoke a host haemocytic reaction (Crompton, 1967). It appears in general that as long as a cystacanth remains intact and healthy, it is protected from a host response, and only wounded or damaged cystacanths which have no envelope are destroyed by the host response (Crompton, 1970, 1975; Lackie, 1972; Lackie & Holt, 1988).

Nevertheless, even healthy and undamaged cystacanths can have deleterious effects upon their hosts. Cystacanths of *Pomphorhynchus laevis* reduce the growth rate of *Gammarus pulex*. They also have a deleterious effect on host respiration. At 20 °C the consumption of oxygen by infected *G. pulex* is reduced by 19.3%, and respiration is also affected at lower temperatures, although this is difficult to detect (Rumpus & Kennedy, 1974). Both sexes are affected in a similar manner. The parasite has an even more drastic effect on egg production (Table 5.2), which is reduced by *c.* 50% in parasitised females. Dezfuli *et al.* (1999) also confirmed that female *Echinogammarus stammeri* infected with *P. laevis* had fewer and smaller eggs. Nevertheless, there appears to be no evidence that male *G. pulex* recognise or selectively avoid mating with infected females even though it would be in their own interests to do so (Poulton & Thompson, 1987). Effects on host

Table 5.2. *Egg production of* Gammarus pulex *infected with* Pomphorhynchus laevis

Locality	Mean ± SE no. of eggs per female		Reference
	Uninfected	Infected	
R. Severn	10.71 ± 0.5	4.82 ± 0.3	Poulton & Thompson (1987)
R. Culm	10.78 ± 0.6	5.53 ± 0.7	Kennedy (unpublished)

growth and reproduction have been reported from other host–parasite combinations. Dezfuli *et al.* (1991b) found that female *Echinorhynchus pungens* infected with cystacanths of *Acanthocephalus clavula* were larger than normal, but never contained eggs. Dezfuli *et al.* (1994) also reported that female *Asellus aquaticus* infected with cystacanths of *Acanthocephalus anguillae* were never observed in amplexus and also had changes to their oostegites. Oetinger & Nickol (1981, 1982) similarly failed to find any cases of amplexus amongst male and female *Caecidotea intermedius* infected with *Acanthocephalus dirus* and also noted alterations of secondary sexual characteristics, suggesting that acanthocephalan larvae induce reproductive disfunction. Brattey (1983, 1986) also reported that larvae of *Acanthocephalus lucii* caused atrophy of the ovaries of *Asellus aquaticus*.

Although there is agreement that acanthocephalans may often affect growth and reproduction of their intermediate hosts, there is far less agreement over whether they cause direct host mortality. Cystacanths are generally found to be overdispersed throughout populations of their intermediate hosts (Table 5.1), and at times the overdispersion may best be described by a truncated negative binomial model (Rumpus, 1973; Amin *et al.*, 1980) suggesting there may be some mortality of heavily infected hosts. Moreover, age-intensity curves are often highly convex (Fig. 5.5). The usual interpretation of these curves is that prevalence and intensity increase with time as the period of exposure to eggs increases. It is also very likely that many infections in small arthropods go undetected because of the small size of the acanthellae, and detection efficiency improves as the larvae develop into the cystacanth stage. However, the decline in parasite abundance in larger specimens is believed to be due primarily, if not exclusively, to selective predation by vertebrates upon infected arthropods (see next section) and so maximum infection levels occur in hosts of intermediate size/age. Any effect the parasite may have on

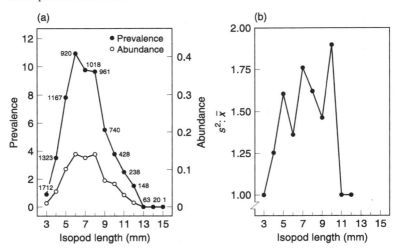

Fig. 5.5 Relationships between length of *Asellus aquaticus* and *Acanthocephalus lucii*: (a) prevalence and abundance and (b) variance to mean ratio in relation to isopod length. Numbers indicate sample sizes. (From Brattey, 1986, with permission.)

host growth will also mean that infection levels will be lower in larger hosts.

Against this background, it is very difficult to determine whether there is any direct mortality due to the parasite. In the laboratory, it has often been found that heavy infection of small individual hosts by eggs may result in their death, but the levels of infection in these studies may well be abnormal and far higher than those encountered in natural populations (Uznanski & Nickol, 1980). However, Brattey (1986) has shown unequivocally that survivorship of *Asellus aquaticus* infected with eggs of *Acanthocephalus lucii* is lower than that of uninfected controls (Fig. 5.6a). By contrast, in a detailed study of laboratory infections of *Leptorhynchoides thecatus* in its intermediate host *Hyalella azteca*, Uznanski & Nickol (1980) found no evidence at all of lethal or sublethal effects of the parasite upon its host (Fig. 5.6b). The parasite did not affect host growth, reproduction or survival even though the distribution of the parasite in the host population could be described by a negative binomial model. They concluded that while death of an abnormally heavily infected host individual could and did occur on occasion, this did not provide evidence of regular direct parasite-induced host mortality. They further suggested that much of the evidence invoked from other studies in support of this possibility was inadequate, as the link between parasites and host death was generally

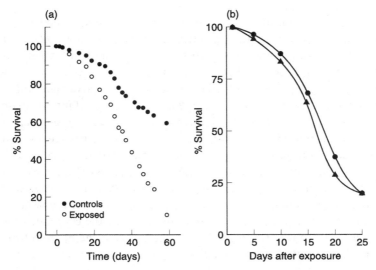

Fig. 5.6 Survival of isopod intermediate hosts experimentally infected with acanthocephalan eggs compared with uninfected controls. (a) Survival of *Asellus aquaticus* containing *Acanthocephalus lucii* (from Brattey, 1986, with permission) and (b) survival of *Hyalella azteca* containing *Leptorhynchoides thecatus*. Solid circles indicate exposed amphipods, triangles indicate controls. (From Uznanski & Nickol, 1980, with permission.)

circumstantial at best and very frequently was unproven. They also pointed out that truncated negative binomial models could be generated by other means, as, for example, through mortality of larger infected individuals as a result of selective predation, i.e. indirect parasite-induced mortality. Overall, it would seem that the case for most species of acanthocephalans *normally* causing direct mortality to their intermediate host has yet to be proven.

5.4 ALTERATIONS IN INTERMEDIATE HOST BEHAVIOUR

5.4.1 In aquatic hosts

In almost every acanthocephalan–intermediate host system studied to date, intermediate hosts infected with cystacanths have shown alterations in behaviour, and in all cases the modifications appear to be adaptive in that they confer greater vulnerability to predation by a potential vertebrate host as a consequence. Although in theory this could be considered an example of host suicide and kin selection (Smith Trail, 1980) where the parasitised invertebrate host sacrifices itself to a predator and thus eliminates the parasite, this model does not seem to be applicable in practice: the changes to host behaviour benefit the parasite but not the host, as the parasite is not taken out of the system. It seems that such behaviour is in fact an acantho-cephalan strategy that improves the probability of a larva being trans-mitted to a definitive host (Moore, 1984). Manipulation of intermediate hosts occurs in species of all three orders of Acanthocephala, in a range of intermediate hosts including insects, crustaceans and ostracods and in aquatic and terrestrial hosts: it would thus appear to be a very old feature of acanthocephalans, having possibly first evolved in the common ancestor of the phylum (Moore, 1984). The behavioural changes occur in normal hosts and in those suffering sterility or other effects of the larvae; the changes only appear when the cystacanth has reached a stage infective to the vertebrate host and the nature of the change is closely related to the foraging and feeding behaviour of the definitive host.

One of the best examples of changes in intermediate host behaviour comes from the studies of Bethel & Holmes (1973). They investigated host behaviour in a series of laboratory experiments in three host—parasite systems: *Polymorphus paradoxus* and *Gammarus lacustris*, *P. marilis* and *G. lacustris*, and *Corynosoma constrictum* and *Hyalella azteca*. A summary of their findings is presented in Table 5.3.

Table 5.3. *The evasive behaviour of* Gammarus lacustris *infected with* Polymorphus paradoxus *and* P. marilis *and* Hyalella azteca *infected with* Corynosoma constrictum

Host and parasite	Mean time ± SE (seconds) in different light zones			
	ULZ	LLZ	Total LZ	DZ
G. lacustris uninfected	16 ± 12	40 ± 24	56 ± 35	3543 ± 35
G. lacustris + acanthor of *P. paradoxus*	42 ± 27	94 ± 83	136 ± 104	3464 ± 104
G. lacustris + cystacanth of *P. paradoxus*	2615 ± 272	534 ± 218	3150 ± 227	450 ± 227
G. lacustris +cystacanth of *P. marilis*	282 ± 79	3318 ± 77	3600 ± 0	0
H. azteca uninfected	0	0	0	3600 ± 0
H. azteca + cystacanth of *C. constrictum*	3570 ± 25	30 ± 25	3600 ± 0	0

ULZ, upper light zone; LLZ, lower light zone; DZ, dark zone.
Source: After Bethel & Holmes (1973).

Uninfected gammarids seek shade and avoid light, and dive for cover on disturbance. Gammarids infected with *P. paradoxus* at the acanthella stage behave in a similar way, but those harbouring infective cystacanths (Bethel & Holmes, 1974) show a positive phototaxis, spending most of their time in the light zone and near the water surface, skimming across the water and then clinging on tightly to aquatic vegetation. The effect of this is to make them conspicuous and to bring them into the surface foraging zone of their definitive bird host. Gammarids infected with the other two species do not exhibit this clinging behaviour, but their normal response to light is completely reversed: *G. lacustris* with *P. marilis* dives on disturbance and remains in the lower parts of the light zone, whereas infected *H. azteca* move into the upper light zone. In a later set of experiments Bethel & Holmes (1977) were able to show that gammarids infected with *P. paradoxus* were more susceptible to predation by ducks and those with *P. marilis* to muskrats. Thus the behavioural changes induced by parasites can differ between conspecifics in such a way as to bring the infected intermediate host into the foraging ambit of the preferred definitive host in a manner specifically adapted to that host.

Parasite-induced changes in gammarids have also been shown to be adaptive to facilitate transmission to fish definitive hosts. *Gammarus pulex* infected with cystacanths of *Pomphorhynchus laevis* shows a similar reversal of its normal evasive behaviour (Kennedy *et al.*, 1978; Brown & Thompson, 1986). Uninfected gammarids show a preference for dark regions of an experimental tank, whereas infected ones are more likely to be found in open water and in the light (Table 5.4). This brings them into the foraging area of their fish hosts, and infected *G. pulex* are selectively preyed upon by a variety of mid-water fish including their definitive host *Leuciscus cephalus* (Tables 5.5, 5.6). Moreover, infected *G. pulex* are also more likely to be found in the drift than uninfected ones (Fig. 5.7), again facilitating predation by mid-water feeding fish species (McCahon *et al.*, 1991). This effect has been confirmed by Maynard *et al.*, 1998), who believe it to result from the preference of infected individuals for illumination and their increased activity. Manipulation can also occur when the gammarid is infected by more than one species of parasite (Cezilly *et al.*, 2000).

Recently, Dezfuli *et al.* (2003) have shown in an experimental arena that infected *Echinogammarus stammeri* were not only more active than uninfected ones but that whereas uninfected ones reduced their activity in response to fish odours in the water, infected ones did not. This failure of infected gammarids to detect the presence of a predator

Table 5.4. *Behaviour of* Gammarus pulex *infected with* Pomphorhynchus laevis

Source	No.	Time (min.)	Mean ($\pm$ SE) no. per min		Significance P
			Light zone	Dark zone	
R. Avon					
Uninfected	5	5	1.35 (0.03) (27%)	3.65 (0.03) (73%)	L v. D <0.001
Infected	5	15	3.61 (0.03) (72.2%)	1.35 (0.02) (27.8%)	L v. D <0.001 U v. I <0.001
R. Severn					
uninfected	20	30	6.87 (0.19) (34.3%)	13.13 (0.37) (65.7%)	L v. D <0.001
infected	20	30	10.53 (0.16) (52.6%)	9.47 (0.15) (47.4%)	L v. D <0.001 U v. I <0.001
R. Culm					
uninfected	5	15	1.46 (0.03) (29.2%)	3.54 (0.03) (70.8%)	L v. D <0.001
infected	5	15	1.54 (0.05) (30.8%)	3.46 (0.05) (69.2%)	L v. D <0.001 U v. I ns

L, light zone; D, dark zone; U, uninfected; I, infected.
Source: Data from Avon and Culm from Kennedy (2003); data from R. Severn from Brown & Thompson (1986).

Table 5.5. *Selective predation by fish upon* Gammarus pulex *infected with* Pomphorhynchus laevis

Fish	Source	*Gammarus pulex* Percentage consumed			Reference
		Inf.	Uninf.	P	
Leuciscus leuciscus	R. Avon	54.0	12.0	**	Kennedy *et al.* (1978)
Thymallus thymallus	R. Avon	43.4	10.0	**	Kennedy *et al.* (1978)
Gobio gobio	R. Severn	76.2	26.2	**	Brown & Thompson (1986)
Rutilus rutilus	R. Severn	85.0	10.0	**	Brown & Thompson (1986)
Alburnus alburnus	R. Severn	50.0	15.0	**	Brown & Thompson (1986)
Leuciscus cephalus	R. Severn	77.0	23.0	**	Brown & Thompson (1986)
Phoxinus phoxinus	R. Culm	55.9	45.0	ns	Kennedy (2003)
Oncorhynchus mykiss	R. Culm	47.0	50.0	ns	Kennedy (2003)

Table 5.6. *Estimates of increase in transmission success from intermediate to definitive host resulting from intermediate host manipulation by an acanthocephalan parasite*

Species of parasite	Intermediate host	Definitive host	Increase in transmission
Acanthocephalus dirus	*Asellus intermedius*	Creek chub	0.27
Pomphorhynchus laevis	*Gammarus pulex*	Gudgeon	0.50
P. laevis	*G. pulex*	3-spine stickleback	0.18
P. laevis	*G. pulex*	Dace	0.58
Polymorphus paradoxus	*Gammarus lacustris*	Duck	0.52
P. minutus	*G. lacustris*	Duck	0.38
Plagiorhynchus cylindraceus	*Armadillidium vulgare*	Starling	0.28

Increase in transmission: difference between the proportion of parasitised hosts eaten by a definitive host and that of unparasitised hosts. All tests were carried out in the laboratory except for that of *P. cylindraceus*.
Source: Data from Bethel & Holmes (1973), Kennedy *et al.* (1978), Camp & Huizinga (1979), Brown & Thompson (1986), Bakker *et al.* (1997), Thomas *et al.* (1998).

would further facilitate the transmission of *P. laevis*. It has also been shown that the colour of *P. laevis* cystacanths, bright orange, assists in attracting some species of predatory fish (Bakker *et al.*, 1997). Lyndon (1996) has also demonstrated the manner in which the behavioural changes of the intermediate host are specifically adapted to the preferred definitive host. *Acanthocephalus anguillae* and *A. lucii* use the same intermediate host, *Asellus aquaticus*, but show different definitive host preferences for *Leuciscus cephalus* and *Perca fluviatilis* respectively. Isopods infected with *A. anguillae* spent twice as long in a light zone as uninfected animals and in response to disturbance, moved towards it: this would tend to increase their susceptibility to predation by *L. cephalus*, which is a disturbance predator. By contrast, isopods infected with *A. lucii* show no significant behaviour changes but instead a pronounced colour change that would render them more susceptible to *P. fluviatilis*, which is a visual predator.

Even the behaviour of larger invertebrates such as crabs may be influenced by the presence of cystacanths of *Profilicolis antarcticus* (Pulgar *et al.*, 1995; Haye & Ojeda, 1998). Latham & Poulin (2002) have investigated the hiding behaviour during low tide of two species of shore crabs infected by *Profilicollis* sp. One species of crab, *Macrophthalmus hirtipes*, was significantly more exposed and had significantly higher infection levels than the other species. Crabs exposed at low tide

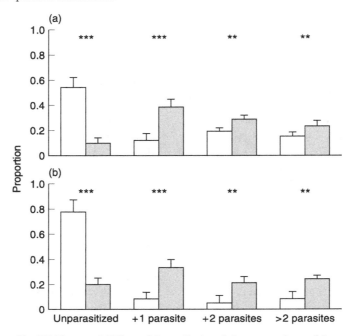

Fig. 5.7 Mean and 95% confidence limits of the proportions of *Gammarus pulex* with different parasite burdens of *Pomphorhynchus laevis* in the drift (shaded) and benthos (unshaded) of the River Teme: (a) margins, (b) mid-river. ∗∗ $P < 0.01$, ∗ $P < 0.05$. (From McCahon *et al.*, 1991, with permission from Blackwell Scientific.)

were held to be at significantly greater risk of predation by definitive shorebird hosts than hidden ones.

5.4.2   In terrestrial hosts

Adaptive behavioural changes have also been reported in terrestrial host–acanthocephalan systems. The acanthocephalan *Plagiorhynchus cylindraceus* alters the behaviour of its intermediate isopod host *Armadillidium vulgare* in an adaptive manner (Moore, 1983a). Infected isopods, especially females, spend more time in the light and less time under shelter than uninfected ones, and they were more frequently found in less humid areas than uninfected ones, rested less often and moved further (Fig. 5.8a,b). These behavioural changes were additional to the sterilisation of females due to the infection. Field experiments showed that the behavioural changes made infected individuals more susceptible to predation by adult starlings, the definitive host, and so to

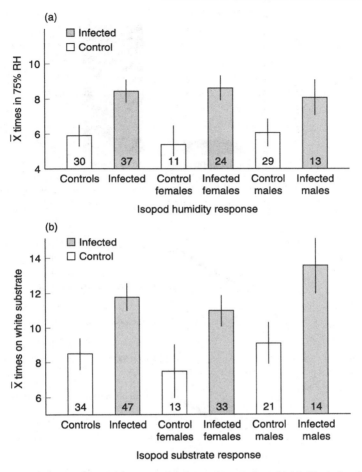

Fig. 5.8 Number of times *Armadillidium vulgare* infected with *Plagiorhynchus cylindraceus* and uninfected isopods were observed in (a) low (75%) relative humidity and (b) on the light substrate half of a behavioural arena. Total observations per isopod = 30 (data from Moore, 1983a). (From Moore & Gotelli, 1990 with permission, Taylor & Francis.)

nestlings which were fed by the parent birds (Fig. 5.9). Moore (1983b) has also been able to demonstrate that the behaviour of cockroaches *Periplaneta americana* is altered by infections with *Moniliformis moniliformis*. Changes have also been observed in other cockroach species, some of which were closely related to *P. americana* and some of which were not (Fig. 5.10), suggesting that the behavioural changes have evolved independently on several occasions. (Carmichael & Moore, 1991; Moore & Gotelli, 1996).

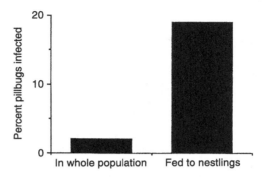

Fig. 5.9 Over-representation of *Plagiorhynchus cylindraceus* in pillbugs fed by starlings to their nestlings (data from Moore, 1983a). (From Keymer & Read, 1991, with permission, Oxford University Press.)

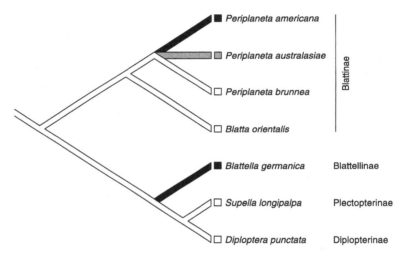

Fig. 5.10 Dendrogram depicting relationships among seven cockroach host taxa and pattern of altered behaviour for use of black horizontal surfaces in relation to infections with *Moniliformis moniliformis*. White, unaltered host behaviour; black, increase in use of black horizontal surfaces by parasitised animals; shaded, decrease in use of black horizontal surfaces by parasitised animals. (From Moore & Gotelli, 1996, with permission.)

5.4.3   Changes in host colour

In addition to altering the behaviour of their intermediate host, many species of acanthocephalan also or alternatively alter the pigmentation of the arthropod. Colouration of arthropods is normally cryptic and adaptive, and this camouflage thus minimises the likelihood of their being preyed upon. There have been numerous reports of the cryptic colouration being partially or completely reversed in arthropods infected with acanthocephalans and these are summarised in Table 5.7. Wherever the effectiveness of this colour reversal has been tested experimentally, it has been shown that the infected arthropods are indeed preyed upon selectively. Pigment changes in three species of isopod in response to infections with *Acanthocephalus* species involved pigmentation dystrophy and the integument of the infected arthropods was significantly more lightly pigmented as measured by determination of light transmission through the opercula (Oetinger & Nickol, 1981). These authors also showed that infection of young isopods resulted in failure of normal pigmentation (Oetinger & Nickol, 1982), but if older individuals became infected their pigmentation remained normal. These changes in pigmentation occurred in addition to deleterious effects upon host reproduction. Lyndon's (1996) study of two species of *Acanthocephalus* in their isopod host *Asellus aquaticus*, (p. 91) showed an increase in pigmentation and melanisation, in the case of *A. lucii* of the respiratory operculae but in the case of *A. anguillae* of the whole body. The differences appeared to relate to different feeding patterns of their definitive hosts, disturbance or visual predation respectively, and were additional to differences in behavioural patterns in infected isopods.

Where arthropods are infected with more than one species of acanthocephalan, there may be a conflict situation. For example, *Gammarus pulex* can be co-infected with *Pomphorhynchus laevis*, which requires a fish host, and *Polymorphus minutus*, which requires a bird host. The vertical distribution of individual gammarids infected with both species was halfway between that expected of each species in a single species infection, but the response to light was dominated by *P. laevis*.

By contrast, Sparkes *et al.* (2004) examined the possible conflict over colour modification between the acanthella stages, which were not infective to a definitive host, and the cystacanth stages, which were, in concurrent infections in the isopod *Caecidotea intermedius*. Host sharing was relatively common in the field, and the stages differed in their effects on hosts when alone. Non-infective acanthellae gave a colour change over 40% of the body, whereas infective cystacanths changed 80%. In conjoint infections, the infective stages dominated (Fig. 5.11).

Table 5.7. *Summary of the effects of acanthocephalan larvae on their intermediate hosts*

Species	Intermediate host	Effects on			Reference
		Colour	Behaviour	Predation	
Palaeacanthocephala					
Acanthocephalus dirus	*Asellus intermedius*	Y	Y	Y	Seidenberg (1973) Camp & Huizinga (1979)
A. jacksoni	*Lirceus lineatus*	Y	Y	P	Muzzall & Rabalais (1975a)
A. anguillae	*Asellus aquaticus*	Y	Y	Y	Lyndon (1996)
A. lucii	*Asellus aquaticus*	Y	Y	Y	Lyndon (1996)
Corynosoma constrictum	*Hyalella azteca*	N	Y	Y	Bethel & Holmes (1977)
Polymorphus minutus	*Gammarus lacustris*	N	Y	Y	Hindsbo (1972)
P. paradoxus	*Gammarus lacustris*	N	Y	Y	Bethel & Holmes (1977)
Pomphorhynchus laevis	*Gammarus pulex*	N	Y	Y	Kennedy *et al.* (1978)
Plagiorhynchus cylindraceus	*Armadillidium vulgare*	Y	Y	Y	Moore (1983a)
Profilicollis sp.	*Macrophthalmus hirtipes*	Y	Y	P	Latham & Poulin (2001), (2002)
Archiacanthocephala					
Moniliformis moniliformis	*Periplaneta americana*	N	Y	P	Moore (1983b)
Eoacanthocephala					
Neoechinorhynchus cylindratus	*Phytocypria pustularia*	N	Y	P	Moore (1984)
Octospiniferoides chandleri	*Cypridopsis vidua*	Y	Y	P	DeMont & Corkum (1982)

Y, demonstrated visually or experimentally; N, not demonstrated; P, not demonstrated but presumed.

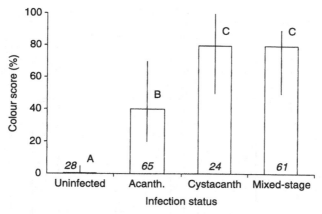

Fig. 5.11 Effect of single and mixed-stage infection of *Acanthocephalus dirus* on colour modification in *Caecidotea intermedius*. Histograms show median values and upper and lower quartiles, with sample sizes in italics. Letters above bars indicate differences between groups, with shared letters indicating no difference. (From Sparkes *et al.*, 2004.)

5.4.4   Failure to alter behaviour

In a very few cases, acanthocephalans have been demonstrated to have no effect upon the behaviour or colouration of their intermediate host. Allely *et al.*, (1992) showed experimentally using standard methods that *Moniliformis moniliformis* had no effect upon the behaviour of the cockroach *Diploptera punctata*. However, this species of cockroach came from the South Pacific region and had not been exposed to the parasite in nature. This was in fact a new association between host and parasite in the laboratory, i.e. there was no evolutionary history of co-adaptation of parasite and host. A similar situation has been reported by Bauer *et al.* (2000). In Burgundy, *Pomphorhynchus laevis* infects *Gammarus pulex*, and alters its behaviour in the manner described previously. Another species, *G.roeseli*, is not native to the area but is a recent coloniser. Uninfected individuals of both species were strongly negatively phototactic: this was reversed for infected *G. pulex* but not for *G. roeseli* (Fig. 5.12). The authors believe that the differential influence of the parasite between host species is the result of the interactions between the parasite's ability to manipulate its hosts and the ability of hosts to resist manipulation. It may be that *P. laevis* is adapted to the native species, but not to the invading one: it can infect the invader, but not modify its behaviour. It is also possible that the invasion process may have selected out resistant individuals of *G. roeseli*.

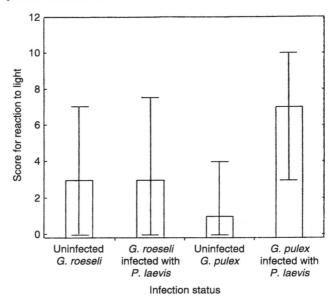

Fig. 5.12 Median values for reaction to light in relation to infection status and identity of intermediate host for a population of *Pomphorhynchus laevis* infecting its normal host *Gammarus pulex* and a new invading species *G. roeseli*. (Reprinted from Bauer *et al.*, 2000, copyright 2000, with permission from Elsevier.)

A further example of failure to manipulate intermediate host behaviour has been provided by Kennedy (2003). This involves *P. laevis* also, but in this case in its normal host *G. pulex*. The normal negative phototactic behaviour of infected gammarids is apparent in the rivers Avon and Severn in England, in which the parasite has been present for a long time (Table 5.4). However, the parasite has only recently appeared in the River Culm, a tributary of the river Exe from which the parasite is absent, being introduced to the Culm with its definitive host *Leuciscus cephalus*. For several years severe pollution in the lower Culm effectively isolated its gammarid population from the Exe, so this population of intermediate hosts and the parasite met for the first time, a situation reminiscent of that in Burgundy above. Similarly, and perhaps therefore not surprisingly, the parasite was able to infect the intermediate host, *G. pulex*, in the Culm, but not to manipulate its behaviour (Table 5.4): here again there is no history of co-adaptation between the parasite and, in this case, its intermediate host population rather than species. Since the parasite shows a different dispersion pattern as well (Table 5.8) such that the *Gammarus* can tolerate higher

Table 5.8. *Frequency distribution of* Pomphorhynchus laevis *in* Gammarus pulex

Source	% of *G. pulex* individuals with *n* parasites per host							N	Effect on host behaviour
	0	1	2	3	4	5	>6		
R. Avon	81.3	13.1	4.3	1.2	0.1	<0.1	0	1685	Positive
R. Culm, Silverton	14.0	29.0	26.0	19.0	9.0	2.0	1.0	100	Negative
Stoke Cannon	17.0	28.0	25.0	21.0	6.0	1.0	2.0	100	Negative

N, number examined.

infection levels in the Culm, it may be that the *G. pulex* population there is more resistant to infections by *P. laevis*.

5.5 SIGNIFICANCE OF BEHAVIOURAL CHANGES IN HOSTS

Overall, the pattern that emerges in respect of acanthocephalan–host relationships is that acanthocephalans in general are not pathogenic to their definitive or paratenic hosts. By contrast, they may have a severe impact on the growth and reproduction of their intermediate host. Almost invariably, they induce host mortality by modifying the behaviour of the host to increase its susceptibility to predation and so improve their probability of transmission to their definitive host. It is certainly true that diseased and moribund animals are rarely observed in natural communities, so the importance of direct parasite-induced host mortality and parasitism as an ecological force has often been underestimated. Nevertheless, it is the indirect effects of parasite-induced intermediate host mortality that is likely to have the major ecological impact.

These parasite-induced changes in behaviour do appear to be truly adaptive. Poulin (1994a,b; 1995) has suggested four conditions for accepting behaviour as adaptive: the patterns must be complex; they should show signs of purposive design; they must increase the fitness of host or parasite; and they are more likely to be adaptive if they have arisen independently in several lineages of host or parasite, i.e. are convergent. Even though many of the behavioural changes are simply increases or decreases in an activity performed by uninfected individuals, there can be no doubt that the parasite is manipulating the changes to its own advantage. Some examples may not meet all four criteria, although many amphipod–acanthocephalan systems do and so, for example, do the *Moniliformis moniliformis*—cockroach and *Plagiorhynchus cylindraceus*—*Armadillidium vulgare* systems: most others seem to meet at least three of the criteria. Other parasite groups provide examples of host manipulation, including cestodes and digeneans (Poulin, 1994 a,b), but in no other group of parasites is it such a universal feature of their host—parasite relationships.

Several authors, e.g. Moore (1983a) and Keymer & Read (1991), have queried why the vertebrates do not avoid infected arthropods. The answer may well lie in the fact that the cost of avoidance may outweigh the cost of infection. Since acanthocephalans are seldom pathogenic in their definitive host (according to Moore & Bell, 1983, even *P. cylindraceus* is harmless to starlings) the extra cost of avoiding infected prey may be prohibitive. Moreover, the energy costs to the vertebrate of chasing and catching prey will be reduced as a consequence of the behavioural changes in the infected arthropod. Manipulation of intermediate hosts may not always result in exclusive transmission to the preferred definitive host, as many predators may feed on the infected arthropods, but this is a trade-off: the cost of some individuals ending up in less suitable hosts is traded against the benefit of others reaching the preferred host.

Life cycles of acanthocephalans depend on a series of unlikely events (Poulin, 1994a,b; 1998) and any adaptations to improve the probability of transmission will be selected for, the more so if there are actually few attendant disadvantages. Behavioural changes in infected invertebrates facilitating parasite transmission may also compensate for the low levels of infection to be found in intermediate host populations (see pp. 19, 156). Dobson (1985, 1988) has shown by the construction of models that enhanced transmission rates will increase the basic rate of parasite reproduction (R_o), which in turn decreases the size of the host population needed to maintain a parasite in its endemic state. Lafferty (1992) has made essentially the same point with his models: weighing the energetic cost of parasitism for the predator against the energetic value of the prey items that transmit the parasite to the predator suggests that there will be no selective pressure to avoid parasitised prey. Acanthocephalans can exploit both predators and prey as hosts, and predators will benefit as long as the energetic costs of harbouring a parasite are low or moderate and the costs of prey capture are reduced. In the words of Dobson (1988), 'acanthocephalans exhibit some of the most highly evolved and sophisticated modifications of host behaviour and these occur in the host with which the parasite has been associated for the longer length of evolutionary time.'

6

Population dynamics

6.1 GENERAL CONSIDERATIONS

The population dynamics of a species can be approached from several different perspectives. One of these is an appreciation that the effects of a parasite upon its host can be considered at several levels: from the effects on an individual host to the effects on a population of hosts and so to the effects on the community in which these hosts exist. Dynamics can also be studied over different timescales, ranging from short-term studies of seasonal changes in population parameters to long-term studies of stability and persistence of parasite populations.

The stability of parasite populations and in particular their ability to regulate their host populations have attracted considerable interest because this can have major consequences for community dynamics. Parasites could thereby affect α and γ diversity and food web organisation. As a consequence, there is a substantial body of theory concerning the operation of regulatory factors. Stability is here considered to be the ability of a parasite population to return to an equilibrium level following perturbation. Stability can only be achieved by the operation of negative feedback controls, i.e. density-dependent factors. The ways in which these factors can operate have been demonstrated in a series of mathematical model systems by Crofton (1971,a,b), Anderson & May (1978, 1979, 1982), May & Anderson (1978) and Dobson & Keymer (1985). In summary, these have shown that a parasite population can be regulated by only a single stabilising factor operating at any stage in the life cycle or in any host, provided that the majority of parasites flow through that host. In a complex life cycle there may be several regulatory factors operating at different stages, and the more there are the greater will be the ability for fine-tuning, but complexity in itself cannot achieve stability.

Anderson & May (1978) and May & Anderson (1978) have focused in particular on the regulation and stability of host—parasite population interactions and the ability of parasites to regulate their own and their host populations. They have shown that three categories of population processes are particularly significant in stabilising and regulating host—parasite systems. These are: overdispersion of parasite numbers per host; non-linear relationships between parasite burden per host and host death rate, i.e. parasite-induced host mortality; and density dependent constraints on parasite population growth within individual hosts. They have similarly identified three categories of population processes that are destabilising. These are asexual reproduction, parasite effects on host reproduction, and time lags in the system. In any particular system, some or all of these processes may be operating and their relative importance will determine whether the system is stable or not. The role of parasite-induced host mortality has sometimes caused some confusion as it has not always been appreciated that for this to be a regulatory factor, the host and its contained parasites must be taken out of the system. Furthermore, as Holmes (1982) has emphasised, mortality must be additive if it is to be regulatory and not compensatory, i.e. it must be additional to other mortality factors and not just replacing one. Compensatory mortality may act as a selective agent in determining which infected hosts will die, but it cannot regulate the system.

There has been considerable discussion over whether parasite populations and parasite—host systems are in fact regulated at all. Price (1980) in particular has emphasised that parasites are adapted to exploit small, discontinuous environments and that they live in non-equilibrium conditions (Fig. 6.1). Their world is a small patch of resources surrounded by other small patches, where the probability of colonising a patch is very low and tenure of a patch is brief. Thus, the loss of any one key feature in a patch could lead to loss of the parasite population. He considers that most parasite—host systems are not stable, that they are probably unregulated and so local extinctions will be common. Anderson & May (1978) and Anderson & Gordon (1982), however, believe that regulatory factors can and do operate on host—parasite systems, and they point out that if they did not, it is highly unlikely that any system could persist over a long period of time. Persistence of a system is not evidence that it is regulated, but is certainly suggestive of the operation of regulatory processes.

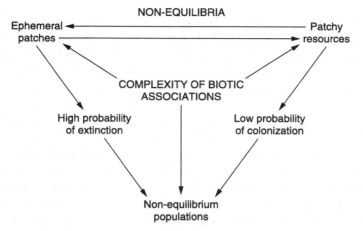

Fig. 6.1 Patch dynamics in parasite populations: the general case. (Reprinted by permission from Peter W. Price, 1980, copyright Princeton University Press.)

All these considerations apply to acanthocephalan populations and to acanthocephalan—host systems. The difficulty, of course, is to determine whether or not the populations and systems are actually regulated in nature. Mathematical models, such as those of Dobson & Keymer (1985) and Dobson (1988), show how regulatory factors could operate and identify the conditions under which these and destabilising factors should operate, but models are only as good as the assumptions made in their construction and are inevitably oversimplifications: they cannot show if the factors identified do actually operate in nature. This requires information from field studies and/or laboratory experiments. Unfortunately, a great many field studies of acanthocephalan populations are essentially short-term studies of seasonal changes. As such, they often fail to provide any evidence of the operation of regulatory or destabilising factors, although they may indicate the existence of such factors in the system. They may also identify correlations between population levels and changes in environmental factors. Experimental studies may indicate that regulatory factors can and do operate, but in many cases the parasite population levels set are far higher than those normally encountered in the wild and their relevance has been questioned for these reasons. The best evidence for regulation can only come from long-term field studies, but as in all branches of ecology, there are far too few of these and often the focus is on one host only. A further problem in relation to acanthocephalans is that most studies on populations have been undertaken on freshwater species in fish and crustacean hosts, and inevitably there can be no certainty that these are representative of marine and terrestrial populations or of acanthocephalans in a variety of bird and mammalian hosts.

6.2 SEASONALITY AND SHORT-TERM STUDIES

The seasonal occurrence of acanthocephalans in freshwater fishes has been reviewed in detail by Chubb (1982). He reported that seasonal cycles in prevalence and abundance of acanthocephalan populations had been demonstrated in representatives of eoacanthocephalans and palaeacanthocephalans in all continents and all biomes; and these cycles were closely correlated with seasonal changes in abiotic environmental factors, especially temperature, and biotic factors, especially host diets.

Over a two-year period, Awachie (1965) studied the periodicity of occurrence of *Echinorhynchus truttae* in both its intermediate host, *Gammarus pulex*, and its definitive host, *Salmo trutta*, in a small stream in Wales. The occurrence of the parasite in its intermediate host showed clear seasonal periodicity. Both prevalence and intensity of infection peaked in mid-winter (January and February) and reached their lowest levels in mid-summer in June and July. Infection levels were higher in larger gammarids, as has generally been found to be the case (pp. 79, 86). Cystacanths were present in the intermediate host, and so available for infection of fish, all through the year. Infection levels in the definitive host also showed a seasonal cycle. Intensity was highest in mid-summer and lowest in winter, and gravid females were found in trout all through the year. Awachie related the cycles to temperature, the life cycle of *G. pulex* and seasonal changes in trout feeding. There were two generations of *G. pulex* a year: a short-lived summer generation and a longer-lived generation that overwintered and reproduced in late spring or early summer. The decline in prevalence and intensity levels in *G. pulex* in summer was related to the death and disappearance of the older, overwintering generation and to the appearance of young and uninfected gammarids. Infection levels in the gammarids were inversely related to infection levels in trout. As water temperatures rose in spring, trout commenced to feed more intensively and so levels of infection increased. Establishment rates were lower at higher water temperatures, but there appeared to be no barrier to superimposed infections (Awachie, 1972). The eggs from these parasites infected the new, overwintering generation. Infection levels increased with fish age, and then declined in the oldest fish which became piscivorous in their diet. Overall, therefore, the cycles related to the breeding cycle of the intermediate host and changes in the diet of the definitive host, both of which in turn were related to water temperature.

A rather similar cycle was reported for *Acanthocephalus lucii* in its intermediate host *Asellus aquaticus* and its definitive host *Perca fluviatilis* in a canal in Scotland by Brattey (1986, 1988), although the seasonality in infection levels was less evident. The isopods showed a seasonal cycle but with a single generation only, breeding in spring and then dying and with a new cohort appearing from June onwards. Infection levels in *A. aquaticus* were lower in summer, reflecting the decline in the overwintering generation of the isopod and the appearance of the new, summer generation. Recruitment of larvae into the isopod population occurred mainly during summer and autumn; and whereas some reached the cystacanth stage by late autumn, others overwintered as acanthellae and developed into cystacanths in spring. Development of larvae could not take place below 6 °C. Two potential regulatory mechanisms were identified: intraspecific competition between parasites in the intermediate host and parasite-induced host mortality. However, population levels were low in the canal and Brattey believed they were below the levels at which these mechanisms operated. In the definitive host, seasonality in the cycle was more evident, with prevalence and abundance being highest in spring and summer and lowest over winter. This related to resumption of feeding by perch as water temperature increased in spring, when the parasites began to mature. Maturation of parasites was related to water temperature, and eggs were shed in late summer and autumn in time to infect the next cohort of *A. aquaticus*. There was thus only a single generation of parasites per year, and although cystacanths were present all year, the cycle related closely to perch feeding behaviour which was in turn related to changes in water temperature. No evidence of density dependent controls was found in the population in fish.

The way in which cycles in one host relate to the other is illustrated in Fig. 6.2. Camp & Huizinga (1980) have reported seasonal cycles in *Acanthocephalus dirus* in both hosts very similar to those reported from *A. lucii* above. They relate cycles in the arthropod to seasonal influx of young arthropods, death of older ones and availability of acanthors, and seasonal cycles in fish to seasonal availability

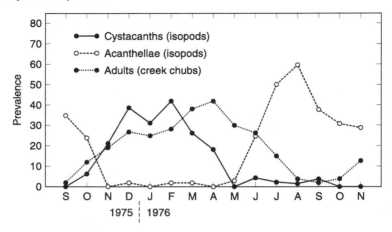

Fig. 6.2 Seasonal variation in the prevalence of infection of *Asellus intermedius* with larvae of *Acanthocephalus dirus* and of creek chubs *Semotilus atromaculatus* with adult *A. dirus*. (From Camp & Huizinga, 1980, with permission.)

of cystacanths and a lifespan of less than a year for adults. In general, seasonal changes in prevalence of acanthocephalans in their definitive host are very common (Fig. 6.3).

Where the intermediate host itself is able to breed throughout the year and shows no evidence of discrete generations, there may be no evidence of seasonality in infection levels in the intermediate host. This is the situation found in *Pomphorhynchus laevis* in *Gammarus pulex* in the warmer waters of southern England (Hine & Kennedy, 1974b), where prevalence and intensity levels show no seasonal pattern. *Gammarus* of all sizes were present throughout the year and individuals harbouring cystacanth stages were present in each month, although prevalence was slightly higher in summer and lower in winter (Fig. 6.4). Adult parasites in dace *Leuciscus leuciscus*, an additional host species, also showed no seasonal cycle in prevalence, abundance or maturation (Fig. 6.4) as dace were able to acquire infections in all months. Experimental infections revealed that establishment rate of cystacanths in fish declined at warmer temperatures (Kennedy, 1972), but was independent of parasite density. Parasite morality in fish was density independent (Kennedy, 1974). What appeared to be happening, therefore, was that a dynamic equilibrium existed between gain and loss of parasites in fish. In winter, fish feeding rates declined at the colder temperatures and fewer *G. pulex* were ingested, but those

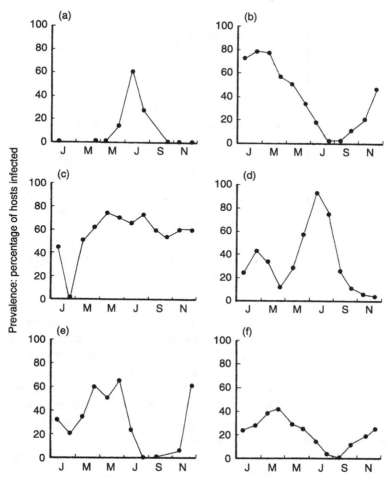

Fig. 6.3 Seasonal changes in the prevalence of (a) *Corynosoma constrictum* in pintail (Buscher, 1965); (b) *Echinorhynchus salmonis* in yellow perch (Tedla & Fernando, 1970); (c) *Neoechinorhynchus saginatus* in fallfish (Muzzall & Bullock, 1978); (d) *Echinorhynchus salmonis* in whitefish (Valtonen, 1980); (e) *Acanthocephalus dirus* in *Lepomis cyanellus* and four other fish species (Muzzall & Rabalais, 1975a); (f) *Acanthocephalus dirus* in creek chub (Camp & Huizinga, 1980). (From Dobson & Keymer, 1985.)

cystacanths that were ingested established better. In summer, fish fed more actively and so ingested more *G. pulex* and *P. laevis*, but the temperature-dependent rejection by the fish meant that fewer parasites established and so infection levels remained fairly similar throughout the year.

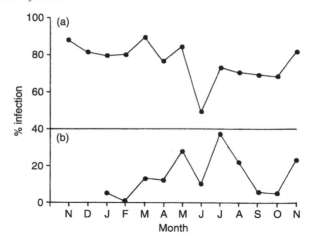

Fig. 6.4 Seasonal changes in the incidence of *Pomphorhynchus laevis*
in the River Avon: (a) adults in dace and (b) cystacanths in *Gammarus pulex*
(Hine & Kennedy, 1974b). (From Kennedy, 1985.)

Brown (1989) studied the population dynamics of *P. laevis* in
both *G. pulex* and in its preferred definitive host *Leuciscus cephalus* in
a different river system, but reached similar conclusions. He found no
seasonal cycle in prevalence or intensity of infections in either
host, maturation of adults showed little evidence of seasonality and
cystacanths were available for infection all the year round. Again,
he postulated a higher turnover in summer and a lower one in winter.
In the River Rhone system, *Gammarus fossarum* is the intermediate
host for *P. laevis*, and although ovigerous females may be found all
year they are most abundant in the summer months, whereas the
prevalence of cystacanths of *P. laevis* peaks in autumn (Van Maren,
1979b).

One of the characteristics of the Irish strain of *P. laevis* is that
it uses *G. duebeni* as its intermediate host. This gammarid exhibits
a pronounced seasonal cycle of growth and reproduction, with the
older generation disappearing in late spring and early summer and
the new generation increasing in numbers in late summer and autumn
(Fitzgerald & Mulcahy, 1983). Although cystacanths can be found in
G. duebeni in all months, the parasite shows seasonality in infection
levels in its definitive host *Salmo trutta*. Both prevalence and intensity
of infection show sharp peaks in May and June respectively, and decline
to lower levels in winter. A far smaller proportion of the adult female
population becomes fully mature in the Irish strain (Molloy *et al.*, 1995)

compared with the English, but the strain is characterised also by seasonality in maturation, prevalence and abundance, associated primarily with the different reproductive cycle of the intermediate host. In North America, *P. bulbocolli* exhibits pronounced seasonality in infection levels in its intermediate host, with levels of prevalence and intensity being highest in winter and lowest in spring. Recruitment into the gammarid host is strictly seasonal and all amphipods appear to acquire their infection at the same time (Gleason, 1987).

Seasonal cycles have also been reported from hosts other than fish and isopods. Helle & Valtonen (1980) have reported seasonality of recruitment of *Corynosoma semerme* into ringed seals *Phoca hispida*, which resulted in a pronounced increase in intensity levels in spring as compared with autumn. However, *C. strumosum* in the same host population remained at a similar intensity in spring and autumn. Seasonality of occurrence of *Profilicollis botulus* has been reported by Liat & Pike (1980) to occur in eider ducks, *Somateria mollisima*. The parasite occurred more frequently and in larger numbers in juvenile ducks and intensities declined with age. Intensity of infection increased with size of the crab *Carcinus maenas*, the intermediate host (Thompson, 1985a), and seasonality in the eiders was not due to changes in the abundance or availability of crabs: although they moved out of the littoral shore zone in winter, they were still available to diving eiders. A substantial proportion of the cystacanth population was located in the larger crabs, but these were infrequently eaten by the ducks, which preferred smaller crabs (Thompson 1985b). Most infections were acquired by young birds early in the first year of their lives, and by the first winter intensities could be as much as ten times higher in these young birds as in adults. The parasites had a short lifespan, and as birds aged there was a natural loss of adults from them; incubating females in particular tend to lose their infections (Thompson, 1985c). They regained them later and the population size was maintained by a regular small intake of infected crabs: once again, a situation of a dynamic equilibrium. The decline in infections in older ducks was attributed to a change in their diet. Epizooitics of *P. botulus* have been reported from time to time and the parasite was thought at one time to cause eider mortality. However, there appear to be no differences in intensities between live and dead eiders (Thompson, 1985c), and Thompson (1985b) suggests that sudden increases in the size of *P. filicollis* populations may occur when eiders are forced to prey selectively on large crabs at times when populations of *Mytilus edulis*, a preferred prey item, have crashed. Population levels appear therefore to be maintained largely by transmission factors, especially the numbers of crabs and the extent to which eiders fed on them, and to be as such unregulated.

6.3 REGULATORY MECHANISMS

All the studies discussed in Section 6.2 have been field based and short-term, with seasonality as their primary aim. Some have provided some information pertinent to regulatory factors almost incidentally (for example, the report by Brattey, 1986, that *Acanthocephalus lucii* does cause direct mortality of *Asellus aquaticus* and the demonstration by Awachie, 1972, that superimposed infections of *Echinorhynchus truttae* are possible in trout) but none has actually demonstrated the operation of a regulatory factor as such. In virtually every acantho-cephalan—intermediate host system there is an indirect parasite-induced host mortality resulting from the effects of the acanthocepha-lan on its intermediate host behaviour, but such mortality is not regulatory. The infected arthropod is not taken out of the system: rather, transmission is facilitated by the mortality. Similarly, many species have effects upon the reproduction of their intermediate host, but parasite effects on host reproduction have been shown to act as a destabilising process (May & Anderson, 1978). There are time lags in the life cycle introduced by the incorporation of an intermediate host and, in some cases, a paratenic host, into the cycle, and if these do have any effect it will again be as a destabilising process. The overall conclusion from these studies is therefore that population levels of acanthocephalans are set by transmission processes, which are often temperature dependent and are certainly density independent in their operation. The short lifespans of many adult acanthocephalans also mean that there is a shortage of time when reproduction is possible and this will set a limit to any population increase.

There are, however, some few studies, mostly from experimental laboratory infections, in which regulatory factors have been identified and have been shown to operate as such. Burlingame and Chandler (1941) undertook a number of experimental infections of rats with differing densities of *Moniliformis dubius*. They showed that in heavy infections, with 40–200 parasites per rat, the percentage survival of the parasites was similar to that in rats given only 20 parasites each. The further survival of the acanthocephlans, however, related closely to their site preference in the intestine. Those that occupied the 'zone of viability' survived, but those forced outside this zone did not.

Thus, intraspecific competition could operate as a regulatory process in this parasite—host system. Furthermore, establishment of secondary infections could be inhibited by the presence of any primary ones. If there was space available in the preferred site, there was no over-crowding and no intraspecific competition for space, so secondary infections could establish and survive. If the preferred space was already filled with parasites or had room only for a few, parasites initially established outside this site but were soon lost from the rat. The numbers that could survive related directly to the size of the optimum site and so competition for space, but within the optimum site, growth of individuals appeared to be unaffected by population density. The size of the preferred site also varies between individual rats.

These experiments were later extended by Andreassen (1975a,b) using greater levels of infection. He found that at levels above 100 cystacanths per rat, the rate of recovery of parasites in a primary infection was very low, in the region of 10—20% after eight weeks. This loss of parasites occurred between weeks four and eight post infection, when establishment fell from 80% to 15%. At lower infection levels, expulsion of parasites occurred later. Expulsion could be prevented by the experimental administration of cortisone to the rats, when 85% of the parasites survived and the parasite density was significantly higher than in control rats. Rats given a heavy primary infection of 100 cystacanths were also dosed with antihelmintic drugs and then challenged with a secondary infection. There was a lower recovery of parasites from the secondary infection compared with the primary, and parasites showed retarded growth and were smaller and lighter than those in a primary infection. They also occupied a more posterior site. Andreassen (1975b) was able to demonstrate the presence of reaginic antibodies in association with the host rejection of the parasites and so he concluded that there was an effective host immune response to the parasites. It is still unclear whether this immune response is additional to the intraspecific competition for space or occurs in association with it, but whichever is the case, there is firm evidence for the operation of a density dependent regulatory process on the infrapopulation size of M. dubius in rats.

Populations of domestic, and probably wild, ducks appear to be able to acquire infections with Polymorphus minutus in all seasons. Seasonality in infection levels may be apparent, and is related to the life cycle of the intermediate host Gammarus pulex and to the seasonal dietary and migratory behaviour of the ducks (Nicholas & Hynes, 1958;

Hynes & Nicholas, 1963). In heavy primary infections there is no decrease in the rate of establishment or in parasite fecundity, although there may be a slight decrease in parasite size. At higher densities of secondary infections, establishment was lower and this was related to the presence of parasites from the primary infection in the preferred site. Individual parasites from secondary infections that did survive in this site grew and bred as normal, whereas those outside the site grew less well and were lost. If ducks were exposed to continual infections, parasites unable to establish in the preferred site were lost and so infrapopulations remained low. Ducks could occasionally be found harbouring high levels of infection, but only if they experienced a massive primary infection were they likely to be harmed or killed. A similar conclusion was reached by Itamies *et al.* (1980). Adult parasites have a short lifespan, up to 50 days at most (Crompton & Whitfield, 1968), and their loss will ensure a regular turnover of parasites. Thus, there appears to exist a regulatory mechanism for this parasite species in ducks very similar to that demonstrated for *Moniliformis dubius* in rats.

Regulatory processes have also been identified in acanthocephalan populations in fish. Brown (1986) infected rainbow trout *Oncorhynchus mykiss* in the laboratory with numbers of cystacanths of *Pomphorhynchus laevis* ranging from 5 to 100 per fish, and monitored their recovery over 12 weeks post infection. Initially the number of parasites establishing related to the infection dose, but at higher levels of infection there was a loss of parasites over the period such that the numbers recovered from infections of 20, 50 and 100 per fish at 12 weeks were very similar (Fig. 6.5). He concluded that his results indicated that in both primary and superimposed infections of *P. laevis* density dependent establishment and survival occurred, such that the number of parasites which survived to reproduce within an individual fish reached a ceiling level. This appears to be a similar process to those described for *M. dubius* and *P. minutus*, even though the preferred site of *P. laevis* in rainbow trout is far less precisely defined. However, Bates & Kennedy (1991a) pointed out that great care had to be taken in interpreting such results as intraspecific competition for space in fish hosts. In their experimental infections of rainbow trout with *P. laevis* the fish grew quite rapidly over a period of four months, the mean intestine length increased significantly and the consequence was that the number of parasites recovered from a constant dosage of 50 cystacanths per fish every four weeks increased from 19.3 to 86.4 (Table 6.1). The distribution of the parasites in the alimentary

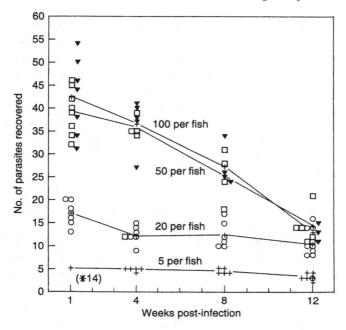

Fig. 6.5 The recovery of *Pomphorhynchus laevis* from *Salmo gairdneri* of
similar size infected with various numbers of cystacanths with time
post-infection. Lines link mean number of parasites recovered with
time for each infection level. (From Brown, 1986, with permission.)

Table 6.1. *Recovery of* Pomphorhynchus laevis *from multiple (50 cystacanths
administered to each fish at 4-weekly intervals) infections of* Oncorhynchus mykiss

	Weeks post-infection				
	1	5	9	13	17
No. of fish	10	7	10	9	10
Mean no. of parasites recovered	19.30	37.14	60.40	78.78	86.40
± SE	1.10	5.10	2.88	6.17	6.32
Range	13−24	20−54	47−75	50−105	60−126

Source: From Bates & Kennedy (1991a).

tract was unaltered, indicating that the preferred site remained
unchanged. There was thus no evidence of a ceiling level of infection
or any other manifestation of competition over 17 weeks, as immigra-
tion of parasites into the infrapopulations always exceeded mortality.

They suggested that the resources, possibly space, were being continually added to and so the limiting resource was only limiting for brief periods. The establishment of secondary and subsequent infections did not therefore depend primarily on the death and loss of a significant proportion of the previously established cohorts.

There have also been difficulties in interpreting results of experimental infection of *Leptorhynchoides thecatus* in both fish and invertebrate hosts. Uznanski & Nickol (1980) compared the survival of the intermediate host *Hyalella azteca* after exposure to parasite eggs with that of uninfected amphipods. They found no difference in survivorship between uninfected and heavily infected individuals (Fig. 5.6b) and concluded that amphipod mortality was independent of cystacanth density. However, when they infected the normal definitive sunfish host *Lepomis cyanellus* in the laboratory with different densities of cystacanths, they found clear evidence of a ceiling in establishment. There was no evidence of a host response to parasites in their preferred site in the intestine, although parenteric individuals were encapsulated and destroyed. At a high density of infection (40 cystacanths per fish) the number of parasites recovered after 14 days declined to levels similar to those in fish fed 20 cystacanths (Fig. 6.6). Following an initial establishment loss of parasites of around 20%, there was no decline in numbers at lower doses of infection.

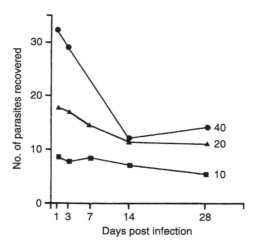

Fig. 6.6 Mean number of *Leptorynchoides thecatus* recovered from *Lepomis cyanellus* fed 10, 20 and 40 cystacanths (Uznanski & Nickol, 1982). From Kennedy, 1985.)

They concluded that establishment and survival of *L. thecatus* in this host were density dependent. The host appeared to be able to provide resource for only around 10–15 adults, i.e. about two parasites per caecum. Below this ceiling, all parasites survived, but above it all in excess were lost. In a later study Ewald & Nickol (1989) confirmed that caecal space was the essential limiting resource. They showed that after three weeks of a similar infection regime, no parasites were found in the anterior portion of the intestine but recovery of parasites from doses of 25 and 40 cystacanths was similar and all parasites were located in the caecae. After an initial preference for particular caecae, by five weeks the 15 survivors were evenly distributed amongst the caecae.

Although *L. cyanellus* is the normal host of *L. thecatus* in Nebraska, in the Great Plains region it uses largemouth bass *Micropterus salmoides* as its definitive host. Leadabrand & Nickol (1993) therefore carried out a similar protocol of experimental laboratory infections with *L. thecatus* in this host. At doses of 10 and 25 parasites per fish, establishment and survival were proportional to the infection dose. At a dose of 40, numbers in the alimentary canal decreased but numbers in parenteral sites increased: parasites here were immature but viable. Caecae did not differ in their suitability. The proportion of cystacanths that established was lower than in sunfish, but unsuccessful parasites at high doses were lost from sunfish but moved to parenteral sites in bass. The numbers surviving in the caecae were similar in the two fish species, but as the bass possessed far more caecae than the sunfish space limitation did not appear to provide a very satisfactory explanation.

A unique example of the identification of regulatory factors in a natural, as opposed to experimental, population of fish was provided by Holmes, *et al.* (1977) and Leong & Holmes (1981). In Cold Lake, *Echinorhynchus salmonis* dominated the lake parasite fauna and infected all 10 species of fish (Tables 4.5, 4.6). Its preferred host was the whitefish, *Coregonus clupeaformis*, which harboured 80% of all gravid females and the flow through which was sufficient to support the whole suprapopulation of the parasite. In this species, the mean number of gravid female parasites per fish remained constant despite an increase in the number of parasites per fish as fish aged and in different seasons. The authors demonstrated a significant negative regression of the percentage of gravid females on the number of acanthocephalans in individual fish, and so concluded that infrapopulations of *E. salmonis* in whitefish were being regulated by density dependent maturation. Infrapopulations in most other host species appeared to be unregulated, with levels determined by transmission factors, although there was a possible regulatory mechanism in lake trout. Regulation in the single species of whitefish would be sufficient to regulate the whole suprapopulation, although other mechanisms may have operated in other fish species.

In all these examples it is possible that any intraspecific competition for space may lead not only to a loss of individuals but also to a decline in average fecundity in the individuals that survive, especially if some are forced out of their preferred niche. This would then function as an additional regulatory mechanism. However, the difficulty of recovering eggs even in experimental infections with acanthocephalans (p. 16) means that this possibility has never been tested. Under field conditions it is difficult to see how this could ever be tested, even when, as in the example of *E. salmonis* above, over-crowding results in a decrease in the number of gravid females.

6.4 LONG-TERM STUDIES

Although the studies reviewed in the previous section provide evidence of regulatory factors that may operate upon acanthocephalan popula-tions, they do not provide any evidence that they do so under field conditions or that populations do in fact remain stable over long periods. Such evidence does, however, exist.

The population changes in *Pomphorhynchus laevis* in both its inter-mediate host *Gammarus pulex* and in a fish host (*Leuciscus leuciscus*) have been recorded in one river over a period of nine years (Kennedy & Rumpus, 1977). Although *L. leuciscus* is not the preferred definitive host it is an additional host and can be considered as a sentinel host for the suprapopulation of the parasite as there is no reason to believe that the proportion of parasites flowing through this host changed throughout the period. Over the period, prevalence and intensity of infection in both amphipod and fish host have remained remarkably constant (Table 6.2). Dobson & Keymer (1985) suggested this constancy may arise because both hosts were present throughout the year and their population densities remained relatively constant at a level deter-mined by factors other than the parasite−predator−prey relationship, i.e. environmental conditions in the regulated river remained constant over the period. By contrast, the later identification of a density dependent factor operating in a fish host of *P. laevis* to set a ceiling level in the number of parasites within an individual fish (Brown, 1986)

Table 6.2. *Long-term changes in the size of the population of* Pomphorhynchus laevis *in the River Avon*

Year	Levels in *Gammarus pulex*		Levels in *Leuciscus leuciscus*	
	Mean %	Mean intensity	Mean %	Mean intensity
1966	No sample taken		74	7.06
1967	19.2	1.5	81	9.6
1968	18.7	1.4	79	9.3
1969	20.1	1.6	72	6.4
1970	19.8	1.3	No sample taken	
1971	16.7	1.5	No sample taken	
1972	17.6	1.5	No sample taken	
1973	17.1	1.4	78	5.1
1974	18.0	1.4	75	4.3

Source: Data from Kennedy & Rumpus (1977).

suggests that the parasite population may in fact truly be stable and that a negative feedback mechanism is operating to produce this situation.

A study of the long-term changes in the population of *Macracanthorhynchus hirudinaceus* in swine in Illinois over a period of 18 years by Dunagan & Miller (1980) has also provided evidence of remarkable constancy in infection levels of the parasite. No information is available on the infection levels in the intermediate host, but variation in prevalence levels in the swine did not exceed 0.5% over the whole period. Even seasonal changes were slight. This constancy may again reflect constancy in host densities and transmission rates throughout the period, but it may equally reflect the operation of an, as yet, unidentified regulatory process; perhaps intraspecific competition for space as has been identified for acanthocephalans in rats and ducks (pp. 110–12).

Other acanthocephalan–host systems provide evidence of population changes around an equilibrium level over time rather than constancy. Parasitic infections in a population of eels *Anguilla anguilla* in one bay in Lough Derg in Ireland have been studied over a period of 18 years by Kennedy & Moriarty (2002). The parasite community was dominated by acanthocephalans, and especially by *Acanthocephalus lucii*. Over the first nine years, annual prevalence levels of *A. lucii* fluctuated irregularly between 78.1% and 97.3%, but over the next six years they declined to a minimum of 32.2% before increasing again to their original level by the conclusion of the study (Table 6.3). The mean abundance of *A. lucii* followed a similar pattern as it declined from 35.5 in the first year of study to below 10 for five years before returning to its original level at 36.2 in the final year. Despite these changes in infection levels, the relative abundance of *A. lucii* varied very little indeed. The overall pattern of changes was suggestive of small oscillations around an equilibrium level, with a severe perturbation being followed by a return to the equilibrium. Infection levels of the other common acanthocephalan, *A. anguillae*, were far more variable and showed an overall decline over the whole period. The decline in infection levels was considered to be due to increasing movements

Table 6.3. *Long-term changes in the population density of* Acanthocephalus lucii *and* Acanthocephalus anguillae *in eels* Anguilla anguilla in *Lough Derg, Ireland*

Year	Acanthocephalus lucii					Acanthocephalus anguillae	
	%	x	SD	s^2/x	p_i	%	p_i
1983	82.6	35.6	28.0	22.1	0.993	12.2	0.004
1884	80.1	35.2	36.5	37.8	0.985	15.9	0.011
1985	88.4	62.0	54.4	47.8	0.984	7.0	0.002
1986	78.1	36.0	32.9	30.2	0.981	8.8	0.003
1987	97.3	37.1	33.1	29.5	0.987	15.7	0.007
1988	88.0	34.8	26.6	20.4	0.971	37.3	0.021
1989	85.8	15.7	9.2	5.4	0.975	14.1	0.020
1990	95.0	56.8	24.0	10.1	0.971	27.5	0.012
1991	87.5	11.5	12.9	14.5	0.972	27.1	0.022
1992	55.8	5.9	11.1	20.7	0.939	2.3	0.023
1993	72.4	4.5	5.6	7.0	0.963	10.3	0.014
1994	60.8	2.2	3.8	6.6	0.982	4.3	0.023
1995	65.8	24.6	22.1	19.8	0.953	15.2	0.037
1996	32.2	2.0	4.0	8.0	0.933	10.2	0.046
1997	66.0	61.0	40.4	26.7	0.978	1.9	0.001
1999	94.7	18.6	21.9	25.7	1.000	0	0
2000	65.0	13.5	20.4	30.7	0.989	5.0	0.007
2001	100.0	36.2	48.1	64.0	0.995	0	0

%, prevalence; x, mean abundance; SD, standard deviation; p_i, proportion of the total number of all intestinal helminth species; s^2/x, variance:mean ratio.
Source: Data from Kennedy & Moriarty (2002).

of eels from other parts of the lough into the study area for unknown reasons. No data on intermediate host populations were available. No regulatory process has yet been demonstrated in this parasite–host system, and if one does exist it may well be operating in the *A. lucii–Perca fluviatilis* system in the lough since the perch there are the preferred host for this parasite. It is, however, difficult to interpret the existence of an equilibrium level and a return to this after perturbation without the operation of a regulatory process.

Other long-term studies on fish hosts have provided some further information on constancy of infection levels. Two studies of parasites of eels in the River Tiber were carried out at an interval of 16 years. In both studies the helminth community was dominated by *Acanthocephalus clavula*: in 1980 it attained a prevalence level of 58.6% and an abundance of 5.6 in summer, and in 1996 the corresponding values were 63.4% and 2.5 (Kennedy *et al.*, 1998). Over the period the relative abundance of the parasite in the community had halved, but the infection levels of eels had remained remarkably similar at the start and end of the period. In other localities population levels of acanthocephalans have fluctuated more erratically and unpredictably. In Neusiedler See the eel population is not natural but has been maintained by stocking. Here, the population level of *A. lucii* has fluctuated rather erratically over an eight year period (Table 6.4) in both sites in the lake, whereas levels of *A. anguillae* have increased over the period and especially at Illmitz. It is possible that such constancy as does exist here reflects changes in eel stocking densities and policies (Schabuss *et al.*, 2005) rather than any regulation.

In a unique long-term study on the acanthocephalans of ringed seals *Phoca hispida* in the Bothnian Bay of the Baltic Sea, Valtonen *et al.* (2004) have provided information on the infection levels of three species of *Corynosoma* over 22 years. It was considered likely that parasite and host population sizes would be coupled, and as seal numbers first decreased markedly over the period and then increased steadily again, so should infection levels in the acanthcephalans. Moreover, the main paratenic host of *C. strumosum* had disappeared from the Bay. In the event, there was no evidence that the mean abundance of any of the three species had changed significantly over time and there appeared to be no relationship between parasite abundance and seal numbers (Fig. 6.7). Populations of *C. semerme* and *C. magdaleni* and even *C. strumosum* remained stable, although this latter species appears more at risk, despite the fluctuating densities of their definite host. All three species shared the same intermediate

Table 6.4. *Long-term changes in the population of* Acanthocephalus lucii *and*
A. anguillae *in eels* Anguilla anguilla *in the Illmitz basin of Neusiedler See*

	Year					
	1994	1995	1996	1997	1998	2001
Acanthocephalus lucii						
Preevalence	21.1	52.6	67.9	54.6	48.8	58.5
Mean abundance	0.8	3.9	8.3	3.7	1.2	6.3
$\pm$ SD	2.4	7.9	18.8	6.4	1.6	9.9
p_i	0.88	0.95	0.87	0.64	0.30	0.39
Variance:mean ratio	7.0	16.2	42.6	11.1	2.0	15.4
Acanthocephalus anguillae						
Prevalence		3.9	21.6	39.4	36.6	80.5
Mean abundance		0.1	0.8	1.1	1.0	9.4
$\pm$ SD		0.3	2.0	1.8	1.9	12.2
p_i		0.01	0.08	0.19	0.24	0.57
Variance:mean ratio		1.5	5.2	3.1	3.6	15.7

SD, standard deviation; p_i, proportion of the total number of all intestinal
helminth species.
Source: From Schabuss *et al.*, (2005).

host *Monoporeia affinis* population, but paratenic hosts were more
variable. In such complex systems as this the links between and coupl-
ing of host density and parasite dynamics are far from simple or
predictable, but the relative constancy of the infection levels of the
three species could suggest a regulatory process in operation.

Overall, it is difficult to obtain a clear picture of the population
dynamics of acanthocephalans. Short-term changes in population
levels appear to be related to density independent seasonal changes
in transmission rates. Laboratory studies have shown the existence of
density dependent processes in several acanthocephalan–host systems
and these could operate to check unusual rises in population levels,
but there is seldom direct evidence that they do so. Mathematical
models (Dobson & Keymer, 1985) have shown how populations could be
regulated and results of the few long-term studies that have been
undertaken suggest that many acanthocephalan populations are
indeed stable over periods of several years. Taken altogether, the
evidence does seem to suggest the possibility that several acanthocephalan populations are in fact stable and regulated around an
equilibrium level.

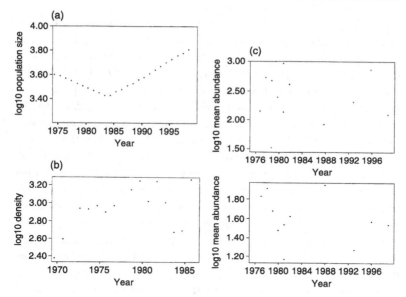

Fig. 6.7 Stability of *Corynosoma* populations in ringed seals over time in the Bothnian Bay of the Baltic Sea. (a) Population size of ringed seals; (b) population densities of the amphipod intermediate host *Monoporeia affinis*; (c) mean annual abundance of *Corynosoma semerme* (top) and *C. strumosum* (bottom) in ringed seals. (From Valtonen *et al.*, 2004.)

6.5 MATHEMATICAL MODELS

The most significant attempt to construct mathematical models of acanthocephalan population dynamics is that of Dobson & Keymer (1985), on which the following account is based. They took as their starting point the basic model of Anderson & May (1978) and May & Anderson (1978) for macroparasites with a simple life cycle and direct transmission and they extended this to encompass the salient features of acanthocephalan life cycles. They focused on the factors controlling the population densities of, and parasite densities in, intermediate and definitive hosts in an attempt to identify the features of the life cycle that were important in maintaining the populations at these densities and how the factors interacted to determine parasite reproductive success.

Dobson & Keymer (1985) assumed as a basis that the host populations had a stable age distribution and a constant size and they considered only cycles with one intermediate and one definitive host.

Parasites were considered to be overdispersed throughout both host populations, and transmission to a host was by ingestion and was governed by the dynamics of predator–prey interactions. They considered that a functional (non-linear) response between egg density and egg ingestion would be likely, as there was some experimental evidence in support of this, but they believed that under the low egg densities likely to be encountered in the wild it would be more valid to assume a direct proportional relationship between density of eggs and hosts. Since little is known about the dynamics of egg survival or cystacanth mortality they assumed a *per capita* mortality rate independent of parasite density or age. Ingestion rates were considered to be directly proportional to densities of intermediate and definitive hosts, although these relationships were complicated by parasite-induced behavioural changes in the intermediate host (Chapter 5, and see also Dobson, 1988, for a more detailed analysis of the consequences of this). The adult parasites would suffer a pre-patent loss due to loss of individual parasites and hosts but in the absence of any data on these rates they assumed an age-dependent survivorship of parasites in vertebrate hosts. They emphasised that values could not yet be assigned to many other key parameters affecting acanthocephalan populations, especially fecundity, which could be affected by host diet, and parasite density-dependent factors such as crowding. They stressed the difficulty in obtaining realistic data on all the necessary life history parameters: data were always scarce and the full complement was unavailable for any species. There were no good data on transmission rates to the intermediate or definitive hosts, on rates of egg production, on survival in relation to parasite density or indeed on egg or adult survival rates.

Despite these difficulties, they were able to construct some generalised models which were inevitably rather theoretical in the absence of detailed estimates of the key parameters. As a starting point, they took the basic reproductive rate, R_0, of Anderson (1982). For the parasite population to persist, R_0 had to equal or exceed 1. It is directly proportional to the density of the definitive host population and, especially, of the intermediate host population and to the lifespan of the adult parasites and of the intermediate hosts (and hence of cystacanths). R_0 is inversely proportional to the cystacanth development time. This means that there are critical host densities below which transmission is impossible and the parasite population cannot persist. Put another way, there is a break point in infrapopulation density when there are too few parasites for the population to reproduce and survive. The parasite population may survive temporary reductions in the size of one host population if the population in the other host is sufficiently high to maintain transmission above the threshold level. The threshold level in the intermediate host is normally satisfied by the high rate of egg production. Moreover, under the right conditions the parasite population can persist when the definitive host is absent for short periods on a regular, e.g. migratory, basis or when it pays regular but occasional visits to the habitat. Fewer definitive than intermediate hosts are needed to maintain the parasite population and more than one combination of intermediate host and parasite population densities will satisfy the equilibrium conditions for zero growth, i.e. it is possible for acanthocephalan parasite densities to attain stable equilibria.

The models suggest that perturbation from the equilibrium will result in oscillations, but these will be damped if the parasite is overdispersed in the definitive host (which is normally the case) and if regulation of the parasite population density occurs (as has been shown to be the case in several systems: pp. 117–20). To satisfy these conditions for stability with oscillations around an equilibrium, the definitive host density at equilibrium should be large, egg production should be intermediate and density of the intermediate host should be low. Low levels of overdispersion in the intermediate host (which is also normally the case) will contribute to the stability of the equilibrium. Under some conditions the parasite population can expand, leading to unbounded growth of the intermediate host population and an increase in parasite numbers to levels determined by density dependent constraints on establishment in the definitive host (as have been shown to operate in *Pomphorhynchus laevis*, *Leptorhynchoides thecatus* and *Moniliformis moniliformis*, pp. 110–15). Changes in temperature or nutrient availability that improve the intermediate host birth rate and so density can be important in determining sudden increases in parasite numbers which will occur when the intermediate host population exceeds a critical threshold. In effect any increase in the values of the parameters which constrain the growth rate of the intermediate host population, including changes in birth, mortality or predation rates, will tend to increase the critical intermediate host population size at which outbreaks to a higher level will occur. Such outbreaks are characteristic of many acanthocephalan populations, notably of species in eiders (Thompson, 1985c; Itamies *et al.*, 1980) and other ducks (Hynes & Nicholas, 1963). High levels of aggregation of parasites in intermediate hosts will also enhance the propensity of parasite and intermediate hosts to break away to upper equilibrial levels.

Widening of specificity at the definitive host level will serve to improve transmission rates in the models. As long as regulation takes place in the preferred host species it can also be viewed as a hedge against local extinction of any species of host. Inclusion of paratenic hosts will complicate the models, but as so little is known about the population parameters of acanthocephalans in these hosts they cannot realistically be incorporated in the models. Their incorporation in a life cycle may be viewed as an adaptation to low or erratic definitive host population densities or to low predation rates by definitive hosts upon intermediate hosts. They require instead high predation rates on the paratenic hosts.

Overall the models suggest that three patterns of acanthocephalan prevalence may be observed:

1. Long-term constancy when there is no evidence of seasonality, where the definitive and intermediate hosts are present all year and where intermediate host densities remain relatively constant due to factors other than parasite—predator—prey relationships. These are evident in *Pomphorhynchus laevis* and *Macracanthorhynchus hirudinaceus* (pp. 116—17)

2. Oscillations, often seasonal, due to seasonal changes in temperature and transmission rates and to a lesser extent on predator—prey relationships, but resulting in stability over the longer term. These are evidenced by *Acanthocephalus lucii* (pp. 117—18).

3. Sudden outbreaks in the populations in definitive hosts, due to expansion of intermediate host populations or changes in the dietary habits of intermediate hosts. These are evidenced by *Profilicollis botulus* and *Polymorphus minutus* (pp. 83, 109).

7

Community dynamics

7.1 GENERAL CONSIDERATIONS

The study of helminth communities is a very recent development in the field of parasite ecology. Holmes (1973) in a seminal paper discussed site selection, site segregation between species, interspecific interactions and the importance of these processes to the development of helminth parasite communities. The subject developed from this point, stimulated by the publications of Price (1980) and publications by Holmes and his co-workers, for example Bush & Holmes (1986a,b), and by the development of mathematical models by Dobson (1985, 1990), Dobson & Roberts (1994) and Dobson & Keymer (1985). The volume by Esch *et al.* (1990) presented in a series of papers the most up-to-date review of parasite community ecology at that time. Since then there have been a series of important contributions from Combes (2001) and Poulin (1998), as well as numerous individual publications by these authors and others such as Bush, Guégan and Kennedy. Acanthocephalans have been included in most of these studies, but generally only as a part of them. Communities comprising only acanthocephalans do occur, but rarely. It is therefore virtually impossible to study the community dynamics of acanthocephalans in isolation in a way that it is possible to study the population dynamics of a species, as the acanthocephalans almost invariably form only a part of any helminth community.

This increase in interest in helminth communities in many respects parallels the recent increase in interest in communty dynamics of free-living organisms. Indeed, parasites present several advantages over free-living organisms for the study of communities. Parasite communities can be studied at a series of nested levels, from infracommunities (a community of parasite infrapopulations in a single host), through component communities (all infrapopulations in a host population or sample) to supracommunities (a community comprising parasite suprapopulations). All helminth community data are actually gathered at the infracommunity level although they may be analysed at any of the three levels. In reality, virtually all studies are undertaken at the infracommunity and component community levels as it is very difficult to obtain the necessary full data sets to analyse communites at the supracommunity level.

Parasite communities are in many respects particularly suitable for the study of community ecology. Because information is gathered at the infracommunity level, it is possible to obtain replicate samples of communities by examining communities in individual hosts of the same species. It is thus possible to obtain quantitative measures of helminth community structure and richness and their variation, and this allows quantitative comparisons of communities to be made between populations of the same host species in different localities and/or between different host species. However, as Holmes & Price (1986) have pointed out, while these features facilitate studies of communities, other features hinder them: in particular the difficulties of defining the resources required by a parasite except in very broad terms of host requirement and site in host. Simberloff (1990) has also stressed the absence of null models from parasite community studies and that, as is also the case in studies of free-living communities, the ghost of competition past is not falsifiable. Nevertheless, both he and other authors such as Cornell & Lawton (1992) have stressed that parasite communities are particularly good places for studying interactions between species. It can be relatively easy to define the niche of a parasite and so study changes in niche use in the presence and absence of other parasite species. Indeed, interspecific interactions may be primary processes in determining the relative abundance of parasite species and therefore in structuring parasite communities (Holmes, 1973; Cornell & Lawton, 1992; Combes, 2001). Many recent studies on helminth communities have focused on the same issues that have attracted ecologists studying free-living communities: for example, whether helminth communities are structured at all and if so, what structures them; whether helminth communities are isolationist in nature or interactive (Holmes, 1973; Bush & Holmes, 1986b); whether communities are saturated or not (Cornell & Lawton, 1992); whether vacant niches exist; whether interspecific competition plays any real part in structuring helminth communities in nature (Holmes & Price, 1986); and what are the relationships between local and regional community richness. The aim of this chapter must therefore be in large measure to determine the role of acanthocephalan species in helminth community structure and functioning.

It is necessary again to stress the limitations on the source and amount of information available to address these questions in respect of acanthocephlans. As stated earlier (p. 31), nothing is known about many acanthocephalan species other than their location and host: quantitative information on their abundance is all too frequently lacking, as is quantitative information on the intensities of other species in the same infracommunities. A large number of studies have other aims, from describing new species to emphasising phylogeny of species, and ecological data are simply not available. Moreover, such data as do exist are heavily biased towards aquatic, and especially freshwater, hosts, simply because it is far easier to obtain large and replicate samples from fish than it is from many mammals. In one respect this may mean that the data that do exist are representative, as it has been stressed in Chapter 2 that acanthocephalans are primarily aquatic, but by contrast it also means that we have very little hard data on the role of acanthocephalans in helminth communities of many terrestrial species of host.

7.2 RICHNESS AND DOMINANCE OF COMMUNITIES

In a search for patterns in helminth community structure and richness in different hosts, Kennedy et al. (1986a) addressed the question of whether there were fundamental differences between communities of helminths in fish and bird hosts. They showed that species richness was indeed greater in birds and some mammals (Fig. 7.1a,b) and mean abundance of helminths and diversity was also higher in these groups. They concluded that the distinction in helminth communities between fish and birds was truly justified and that the differences were valid and fundamental. They identified the factors responsible for diversity in helminth communities as being: complexity of the alimentary canal; host vagility; a broad host diet; selective feeding by the host on prey which serve as intermediate hosts for a wide variety of helminths; and exposure of hosts to direct life-cycle helminths. In a later paper, Bush et al. (1990) focused on the role of ecological and phylogenetic determinants of helminth community richness, and showed that terrestrial hosts had on average species

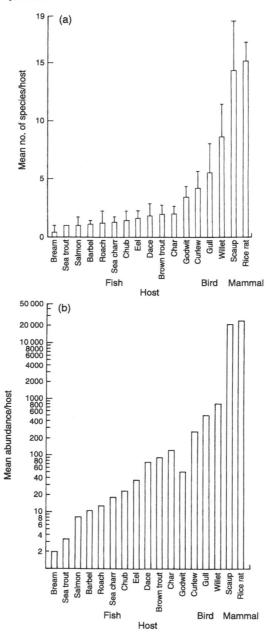

Fig. 7.1 Patterns in intestinal helminth communities in host species. (a) Comparative mean species richness and (b) comparative mean abundance. Data for fish represent the most diverse site or highest values for each species. Bars denote ± 1 SD. (From Kennedy *et al.*, 1986a.)

poorer communities than aquatic hosts. The richest intestinal communities in terrestrial hosts were found in mammals, but the richest communities in aquatic hosts were found in birds, and they concluded that habitat of the host was more important in determining community richness than host phylogeny. Although fishes, as the oldest group, might be predicted to have the richest helminth communities, they did in fact have the poorest.

These richer communities in birds and mammals tend to lie at the interactive end of the isolationist–interactive continuum (Holmes & Price, 1986). They are species rich, their structure is deterministic, it is possible to identify core and satellite species within the community, and there are few or no vacant niches (Table 7.1). Much of the richness of parasite communities in terrestrial mammals is due to the presence and abundance of cestodes and nematodes with direct life cycles (see examples in Pence, 1990, and Lotz & Font, 1994): acanthocephalans may be present in some communities in some localities but they are seldom if ever dominant, they are almost invariably satellite species and they appear to play only a minor part, if any, in structuring these communities. Similar considerations

Table 7.1. *Correlates of isolationist and interactive communities*

	Isolationist⟵⟶ Interactive	
Composition	Species poor	Species rich
	Stochastic	Deterministic
	No core species	Core species
	Vacant niches	No vacant niches
Structure	Unstructured	More structured
	Unsaturated	Saturated
	Low population sizes	High population sizes
	Low p of colonisation	High p of colonisation
	Non-equilibrial	Equilibrial
Interspecific	Low p of interactions	Higher p of interactions
competition	Species assort independently	Species sensitive to others
	Species individually dispersed	Species evenly dispersed
	Unimportant	Important
Site dependency	Individual adaptation	Interactions
	or p of mating	

p, Probability.
Source: Based on ideas presented in Holmes & Price (1986) and Holmes (1987).

apply to helminth communities in birds. The rich communities of most species of aquatic birds appear to be interactive and dominated by cestodes and digeneans (Avery, 1969; Bush & Holmes, 1986a,b; Edwards & Bush, 1989; Stock & Holmes, 1988; Goater & Bush, 1988; Bush, 1990), and although acanthocephalans are often present, and in some localities abundant (Thompson, 1985a,b), they appear normally to be relatively minor components with, in general, relatively low levels of prevalence and abundance. Thus in these groups acanthocephalans often contribute to species richness but have little impact on community structure or diversity. Helminth communities of amphibians and reptiles are often depauperate and highly variable in composition and have features characteristic of isolationist communities (Aho, 1990), with the possible exception of communities in turtles. These may be species rich, and harbour up to four congeneric species of *Neoechinorhynchus* (Aho *et al.*, 1992; Aznar *et al.*, 1998), but it has yet to be demonstrated whether the communities are also interactive.

It might be predicted therefore that acanthocephalans will only play a major role in community composition and structuring in hosts in which they are abundant or even dominant and in which helminth communities are species poor and isolationist in nature. This would include seals, in which intestinal communities are frequently dominated by acanthocephalans, especially species of *Corynosoma* (Helle & Valtonen, 1980, 1981; Valtonen & Helle, 1988; Nickol *et al.*, 2002b; Valtonen *et al.*, 2004). There is no doubt that acanthocephalan species are the major component of the helminth communities in these hosts, but it is not yet clear whether the communities are actually structured by their interactions or whether they are isolationist in nature. Acanthocephalans may also be a major component in the helminth communities of whales. Aznar *et al.* (1994) have shown that *Polymorphus cetacea* is the dominant species in the intestinal helminth communities of *Pontoporia blainvillei*, accounting for 99% of all individual helminths found and with a mean abundance of 397 per host. These communities are depauperate (only five species) and predictable due to the regular occurrence of the acanthocephalans but the infracommunities are considered to be stochastic in occurrence and isolationist in nature. Helminth communities in *Globicephala melas* are similar in several ways, but species abundance levels are higher and the predictability of community composition even stronger (Balbuena & Raga, 1993).

It could be predicted that acanthocephalans might play a more important role in helminth communities of fish, partly because they are themselves an aquatic group and partly because helminth communities in fish are often species poor and isolationist in nature (Kennedy, 1990). There is little evidence that they are important in structuring helminth communities in marine fish (Holmes, 1990), in which communities are generally dominated by digeneans. However, this may reflect the paucity of studies specifically directed towards analysing community structure in marine teleosts. Where one might expect to find acanthocephalans playing a major role in community composition and structure is in the intestinal helminth communities of freshwater fish and especially in temperate regions of the globe where acanthocephalans are prominent or even dominant members of the communities (pp. 33−7). The concept of core and satellite species is not really applicable to these communities as so many are actually species poor.

The most detailed studies of helminth communities of fish have been carried out in northern temperate regions and on several species including eels *Anguilla anguilla*, brown trout *Salmo trutta* and some other species including barbel *Barbus barbus*.

The occurrence of multiple congeneric species is uncommon in fish hosts and among acanthocephalans (Fig. 3.1). Indeed, species flocks as such are unknown in this group and single species infections are by far the most common; it is not very common to find even a single pair of congeners in a helminth community in fish.

It is also often found to be the case that not only is there only one species of acanthocephalan in a particular host species, but also there is only a single species in any particular locality, or one species that dominates that locality (Table 7.2). This may often be due to the distributions of intermediate hosts: for example, in western Europe *Acanthocephalus clavula* uses *Asellus meridianus* as its intermediate host whereas *A. anguillae* and *A. lucii* use *Asellus aquaticus*. Since the two species of *Asellus* compete they rarely, and then only fleetingly, co-exist in the same locality and so it is very unusual to find *A. clavula* and any other species of *Acanthocephalus* in the same locality. It is also very unlikely that *A. anguillae* and *Pomphorhynchus laevis* will co-occur in a locality since the two species compete (pp. 152−3). Such dominance of all host species by a single species of acanthocephalan is also possible because of the ability of acanthocephalans to utilise a wide variety of host species.

Table 7.2. *The occurrence and dominance of acanthocephalans in fish in British waters*

Site	Proportion of each species in all fish species					
	A. clavula	A. lucii	A. anguillae	P. laevis	E. truttae	N. rutili
R. Clyst	1					
R. Avon				0.99	0.005	0.005
R. Stour (Suffolk)	1					
R. Ouse	1					
R. Elan						1
R. Severn				1		
R. Stour (Dorset)				1		
R. Terrig					1	
R. Alyn					1	
R. Dee	1					
R. Trent			1			
S. Union Canal		1				
Forth & Clyde Canal		1				
Leeds/Liverpool Canal		1				
L. Bala	0.97				0.02	0.01
L. Celyn					0.03	0.97
Hanningfield L.						1
L. Leven					0.88	0.22
L. Padarn	0.77				0.01	0.22
Rostherne Mere		1				
Serpentine		1				

Data are expressed as the proportion of the total number of acanthocephalans found that are due to each species, and results from all species of fish are combined.
Source: From Kennedy (1985).

In the British Isles there are only six species of acanthocephalans that infect freshwater fish and all six species are known to occur in eels (Table 7.3). Some of these are more commonly found in eels than others but all, with the possible exception of *Echinorhynchus truttae* which is a specialist of salmonids, are able to mature in eels. Between them, they dominated the majority of both component communities and infracomunities (Table 7.3 and Fig. 7.2) and when present in a community they almost invariably dominated it. The communities were on the whole depauperate with a normal maximum number of species per infracommunity of three (Kennedy, 1990).

Table 7.3. *The occurrence and dominance of acanthocephalan species in the intestinal helminth communities of eels* Anguilla anguilla

	A. clav.	A. lucii	A. ang.	P. laev.	E. trut.	N. rut.
Component level						
No. of sites where it occurs (/50)	13	20	9	5	2	3
No. of sites where dominant (%)	10.6	31.9	4.2	4.2	4.2	2.1
Sites in which it occurs where dominant (%)	38.5	75.0	22.2	40.0	100.0	33.3
Infracommunity level						
No. of eels where it occurs ($n = 842$)	101	127	32	17	31	3
Infected eels where dominant (%)	21.1	22.9	4.8	3.9	6.9	0.7
Eels in which it occurs where dominant (%)	90.1	77.9	65.6	100.0	96.7	100.0

A. clav., *Acanthocephalus clavula*; A. ang., *A. anguillae*; P. laev., *Pomphorhynchus laevis*; E. trut., *Echinorhynchus truttae*; N. rut., *Neoechinorhynchus rutili*.
Source: Modified from Kennedy (1990), with kind permission of Springer Science and Business Media.

They were considered to be isolationist in character with species being stochastic in occurrence. When, however, the relationship between regional helminth species richness and local species richness was investigated for 32 species of freshwater fish in Britain, Kennedy & Guégan (1994) found it to be curvilinear in shape for native species (Fig. 7.3), indicating that component communities became saturated well below the level of regional species richness: for eels regional species richness was 20, but maximum local species richness was only nine. This could reflect merely a supply-side situation, or could be evidence of structuring in the community. Whichever is the case, it appeared that local patterns in community composition were just that: they were unaffected by regional patterns and these could not be used to predict local community composition. In respect of the acanthocephalans, this confirmed that any one of the six species could be present and dominate an eel component community and which one could only be predicted,

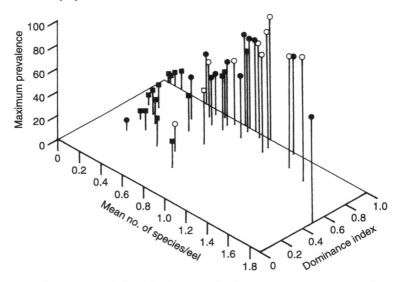

Fig. 7.2 Interrelationships between the three summary parameters of helminth component communities in eels *Anguilla anguilla* in 50 localities from the British Isles. Each point represents one locality. Solid symbols, river; hollow symbols, lake; circles, dominance by acanthocephalan generalists; squares, dominance by eel specialists. (From Kennedy, 1990, with kind permission of Springer Science and Business Media.)

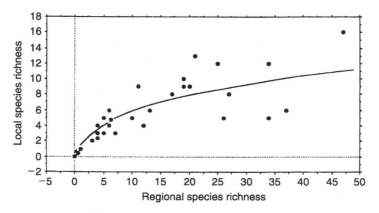

Fig. 7.3 Relationship between the number of regional helminth species and local helminth species for each of the total 32 species of freshwater fish in the British Isles. (From Kennedy & Guégan, 1994.)

if at all, on the basis of local characteristics such as other species of fish present in a locality and particular species of isopod or amphipod intermediate host available.

Kennedy & Guégan (1996) then went on to focus on the number of niches in intestinal helminth communities of eels. By examining the number of species found in infracommunities in 64 localities and in a single locality over a period of 17 years in the British Isles they found that the maximum number of species harboured by an individual eel was four (Table 7.4). The great majority of eels harboured only one species or were uninfected. They also found that the relationship between infracommunity richness and component community richness was curvilinear and so infracommunities were saturated below the component community richness level. They considered that this limitation on the number of species could not be explained by supply-side ecology, pool exhaustion or transmission rates. Rather, they believed that there must be infracommunity processes acting to impose the limit. Most infracommunities had one species of acanthocephalan (any of the six) and the most frequent community composition was a species of acanthocephalan and one or two other species from different taxa. They further suggested that competition between species, involving the acanthocephalan, could be operating even in a species poor community.

Table 7.4. *Frequency distributions of the number of intestinal helminth species per eel* Anguilla anguilla *for England and Ireland separately and combined, for the River Clyst and for* Anguilla reinhardtii *in Australia*

		Number of helminth species per eel							
		0	1	2	3	4	5	6	7
British Isles	No. of eels	629	432	164	49	3	0	0	0
	%	49.3	33.8	12.8	3.8	0.1	0	0	0
Ireland	No. of eels	126	115	39	2	0	0	0	0
	%	44.7	40.7	13.8	0.7	0	0	0	0
England	No. of eels	503	317	125	49	0	0	0	0
	%	51.0	32.0	12.5	4.5	0	0	0	0
R. Clyst	No. of eels	732	351	72	17	3	0	0	0
	%	62.3	29.8	6.1	1.4	0.2	0	0	0
Australia	No. of eels	10	24	23	28	17	11	2	1
	%	8.6	20.7	19.8	24.1	14.6	9.5	1.7	0.9

Source: From Kennedy & Guégan, (1996).

Other workers have disagreed with their conclusions. Rohde (1998) suggested that the curvilinear relationship was the consequence of different likelihoods of parasite species appearing in the infracommunity as determined by transmission rates and lifespans. Processes operating at the infracommunity level were therefore not necessary to explain the curvilinear relationship and he believed that there was no need to invoke interspecific competition as an explanation (Rohde, 1994). More recent studies by Norton et al. (2003, 2004a,b) on two rivers where four and three species of acanthocephalans have been found in helminth infracommunities in eels have suggested that the helminth communities in eels are in fact unsaturated and that their composition reflects that of the local component community. The dominance in any component community in turn reflected the local fish and invertebrate fauna. The authors did feel that the two component communities studied were unlike those reported from other parts of Europe in that both component (three and four species of acanthocephlans) and infracommunity (most frequent category was two species per eel, not zero or one) richness were higher than in Europe (Conneely & McCarthy, 1986; Callaghan & McCarthy, 1996; Schabuss et al., 1997, 2005; Kennedy et al., 1998; Borgsteede et al., 1999) where communities exhibited lower diversity. Norton et al. (2004a) also reported temporal and spatial patterns of nestedness in species composition in intestinal helminth component communities of eels, particularly when communities exhibited high dominance by a single species and not when species abundance was similar, and they considered that this reflected a colonisation process. Overall, there is agreement that helminth infracommunities in eels are often dominated by a species of acanthocephalan and that the communities are in general species poor, but the question of whether they are isolationist in nature or structured by interactions between species remains unresolved (see also pp. 153–4)

Helminth communities in other species of freshwater fish share many characteristics in common with those of eels. Kennedy & Hartvigsen (2000) investigated the component and infracommunities of brown trout from 72 localities in Britain and Norway. Communities in trout were more predictable than those in eels, as a recurrent group of four species of trout specialists gave greater predictability to the community composition. In most localities a trout specialist dominated the community rather than an acanthocephalan generalist as in eels, but a species of acanthocephalan dominated 55% of the component communities in Britain and 36.4% of the communities in Britain and Norway combined, with the specialist *Echinorhynchus truttae* being the single most common (22.2%) dominant. The relationship between maximum infracommunity richness and component community richness was curvilinear, suggesting that infracommunities were saturated below levels of component community richness. They found that all measures of community structure and indices of richness and diversity indicated that helminth communities in trout were isolationist in character, species poor and exhibited low diversity at both community levels. In fact, all values for trout helminth communities were strikingly similar to those obtained from eels; a limit of four species per infracommunity further enhanced the similarity and suggested a common determinant of community structure. In Ireland, where the fish parasite fauna as a whole is less rich than that of fish in Britain, the great majority of helminth communities at both levels in brown trout are dominated by a species of acanthocephalan; this can be any one of the six species found in Britain but is commonly *Pomphorhynchus laevis* (Molloy *et al.*, 1995; Byrne *et al.*, 2000, 2002, 2003, 2004).

There are few comparable studies on other species of fish, but one exception is the barbel. Almost everywhere that barbel are common in a river, for example the Danube basin, the River Tiber, the River Rhone and the Rivers Severn and Avon (Table 3.9), both infracommunities and component communities are heavily dominated by *Pomphorhynchus laevis*.

7.3 SPATIAL AND TEMPORAL VARIABILITY

If, as has been suggested, helminth component and infracommunities in freshwater fish are depauperate in species and isolationist in character, then it might be predicted that there will be very little similarity between communities in space and time. Even dominance of a community by an acanthocephalan species will not of necessity increase similarity since all the acanthocephalans are to a greater or lesser extent generalists, any one of which is capable of infecting a wide variety of fish species in different localities. Even the overwhelming dominance of barbel intestinal communities by *Pomphorhynchus laevis* does not ensure high levels of similarity between communities (De Liberato *et al.*, 2002). The highest levels of similarity were found between communities in the same river, the Danube, and levels of similarity between communities in the Danube

Table 7.5. *Sorensen's index of similarity between intestinal faunas in* Barbus *spp.*
from some European localities

| | B. tiberinus | B. barbus | | | | B. cyclolepis | B. b. bocagei |
		(A)	(B)	(C)	(D)		
Barbus tyberinus	—						
Barbus barbus (A)	0.43						
Barbus barbus (B)	0.40	0.61					
Barbus barbus (C)	0.44	0.57	0.89				
Barbus barbus (D)	0.30	0.35	0.60	0.67			
Barbus cyclolepis	0.40	0.26	0.40	0.44	0.30		
Barbus b. bocagei	0.35	0.10	0.12	0.13	0.12	0.35	—

B. tyberinus from R. Tiber, near Rome, Italy; *B. barbus* (A) from R. Danube near
Budapest, Hungary; *B. barbus* (B) from R. Danube near Vienna, Austria; *B. barbus*
(C) from Ischler Ache Brook, Austria; *B. barbus* (D) from R. Jihlava, Czech Republic;
B. cyclolepis from R. Maritsa, Bulgaria; *B. b. bocagei* from north-western Spain.
Source: From De Liberato *et al.* (2002), with permission.

and in other rivers did not exceed 0.44 and were generally far
lower than this (Table 7.5). The presence of specialists dominating
helminth communities in trout may make them more predictable
(Kennedy & Hartvigsen, 2000), but there are no data available to
show whether they are consequently more similar. A comparison of
communities in eels from four neighbouring localities in Belgium
by Schabuss *et al.* (1997) is illustrative of the extent of local variation: acanthocephalans were absent from two of the localities while
Acanthocephalus lucii and *A. anguillae* were present in the other two,
of which one was dominated by *A. lucii* and the other by *A. anguillae*.
Carney & Dick (2000) found a similar situation when they studied
helminth communities of yellow perch *Perca flavescens*. They reported
three species of acanthocephalan from five lakes, but *Neoechinorhynchus*
sp. occurred only in one lake at low prevalence and abundance
and *Pomphorhynchus bulbocolli* occurred in two lakes at low prevalence
and abundance. *Echinorhynchus salmonis* was reported from only one
lake, Lake Michigan, in which it appeared to be a significant member
of the community but did not dominate it. Not surprisingly, they
concluded that patterns in species occurrence were not predictable.

Poulin & Morand (1999) suggested distance between conspecific
host populations would be a key determinant of the likelihood that

exchanges of helminth species would occur between these populations and that this variable would therefore influence the similarity between helminth communities. Using data sets on component communities of helminths in brown trout obtained by Kennedy (1978b), they showed that geographical distance between localities was a predictor of similarity. However, using another data set on helminth communities in brown trout in a number of reservoirs in south-west England (Kennedy et al., 1991; Hartvigsen & Kennedy, 1993) they confirmed the findings of these authors that similarity in species richness actually increased with increasing distances between reservoirs and was not a good predictor of similarity. In many cases inspection of data sets without formal analysis suggests that distance may be unimportant: helminth communities in eels in the River Avon are dominated by *Pomphorhynchus laevis* (Table 3.9), but those in the River Test, a neighbouring river, are dominated by *Echinorhynchus truttae*. This species occurs at low levels in the River Avon (Hine & Kennedy, 1974a) but *P. laevis* is not found at all in the River Test (Norton et al., 2003). Similarly, an inspection of their comparison of eel component community diversity between England and Germany (Table 7.6) reveals considerable differences in the identity of the dominant species, the species of acanthocephalan that dominates, and measures of species richness and diversity.

Table 7.6. *Comparison of helminth component communities dominated by acanthocephalans between eels* Anguilla anguilla *from selected European localities*

	NHS	H'	Si	BP	Ds	%In.	%0/1
England							
R. Test[a]	8	1.92	2.72	0.54	E.t.	86	38
R. Clyst[b]	3	1.04	2.67	0.58	A.c.	30	99
Germany							
R. Rhine – Alb[c]	6	1.11	2.12	0.67	P.a.	58	63
R. Rhine – Worms[c]	4	0.42	1.23	0.90	P.a.	20	95

NHS, number of helminth species; H', Shannon diversity index; Si, Simpson's diversity index (reciprocal); BP, Berger-Parker dominance index; Ds, dominant species; %In, % of eels infected; %0/1, % of eels infected with 0 or 1 species.

E.t., *Echinorhynchus truttae*; *A.c.*, *Acanthocephalus clavula*; *P.a.*, *Paratenuisentis ambiguus*.

Source: Data from Norton et al. (2004b)[a], Kennedy (1993b)[b], Sures & Streit (2001)[c].

Poulin (2003) returned to this theme of the predicted decay of community similarity with geographical distance in helminth communities of vertebrate hosts. Using populations of three species of fish and three of mammals in North America, he calculated the similarity between communities for all possible species pairs. He predicted that similarity would decay with distance at an exponential rate, and found the prediction to be verified for two species of mammal and two of fish. However, the rate of decay appeared to bear no relationship to host vagility. Elsewhere, in Mexico, Vidal-Martinez and Poulin (2003) have addressed the same question of replicability in helminth communities. They studied one species of marine fish *Epinephelus morio* from nine localities and one species of freshwater fish *Cichlosoma urophthalmus* from six localities and from two localities over a period of a year. They looked particularly for evidence of nested patterns and examined associations between pairs of common species. Positive associations and nested patterns were noted in some localities for both species and at some sampling times, but non-random patterns were only observed sporadically. They predicted that adjacent localities were more likely to display similar helminth community structure than distant ones and so distance would be an important determinant of predictability and similarity. They found some associations in the data set, but overall no patterns were evident. There was no consistency in the subgroups exhibiting nestedness among or between the localities. Patterns from one site were not representative of others, and even when departures from randomness occurred they were not repeatable in time or space. Acanthocephalan species were present in the data set, but played only a minor role in their communities.

The few long-term data sets that are available also suggest that communities are not going to be very similar. An analysis of helminth communities in eels in one river over 19 years (Kennedy, 1993b) showed pronounced changes in component community composition and structure (Table 7.7). For the first four years, the community was overwhelmingly dominated by *Acanthocephalus clavula*, but this species declined in 1983 when the dominant species was a cestode. From 1987 onwards, the dominant species was a nematode, although *A. clavula* was generally present at a low level and *Neoechinorhynchus rutili* had appeared. Throughout the period species richness fell from three to zero, then peaked at nine and varied erratically thereafter. Similarity of communities between 1979 and 1998 was only 22% and between 1983 and 1987 only 11%. Variation in

Table 7.7. *Changes in the relative abundance (as a proportion* (p_i) *of the total number of all helminths of all species) of acanthocephalans in the intestinal helminth component community of eels* Anguilla anguilla *in the River Clyst*

	1979	1980	1981	1982	1983	1984	1987	1991	1994	1995
No. of species	3	3	3	3	2	0	9	9	4	7
% infected	69.3	56.1	30.2	15.9	11.6	0	56.2	58.4	58.8	80.9
A. clavula (p_i)	0.78	0.88	0.5	0.74	0.25	0	0.01	0.02	0.27	0.03
N. rutili (p_i)	0	0	0	0	0	0	0.01	0.01	0	0.08

Source: Modified from Kennedy (1993b).

composition and dominance was also observed over 13 years in the nearby River Otter (Kennedy, 1997a). In 1985 the community was species poor but within three years it had changed to being quite rich (3–8 species) (see Table 8.8). Initially only *N. rutili* was present, but two years later *A. clavula* and *Pomphorhynchus laevis* had appeared: levels of this latter species increased throughout the period but the other two species had disappeared by the end of the investigation. The overall conclusion of all these investigations must therefore be that there is a high degree of both spatial and temporal variation in the composition of helminth communities in freshwater fish, and the presence and role of acanthocephalans in their composition, richness and structuring is also variable in place and time and is seldom, if ever, predictable.

7.4 NICHE DYNAMICS

It is sometimes tempting to believe that there exist almost as many definitions of a niche as there are ecologists. However, rather than attempt to define a niche as such, it may be more valuable to try and describe it. The niche of an organism can be regarded as its range in ecological rather than geographical space (Whitfield, 1979). For any particular species, the niche represents the optimal concentration of the essential environmental resources that the species requires. Where resources are arranged and change along a longitudinal axis, as is the case in the vertebrate alimentary canal, each parasite species will distribute itself along the alimentary canal in relation to its resource requirements (Fig. 7.4). The principal resources available to helminths along the alimentary tract can be

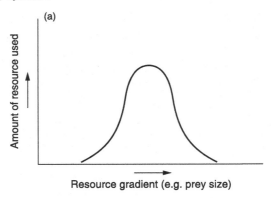

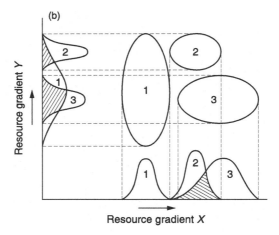

Fig. 7.4 Niche utilisation. (a) Diagram of a resource utilisation curve, demonstrating the differential use of a resource over (in this example) a range of sizes. (b) A segregated array of resource utilisation curves for three species along a resource gradient, for example an intestinal tract. Each species shows a preference for a particular region, but their distributions nevertheless overlap. (From Whitfield, 1979, p. 216, with permission, copyright Hodder Education.)

summarised as nutrients and space. In the case of acanthocephalans, it is not yet known what specific nutrient resources they require or the extent to which different species may require different resources, but it is possible to describe their use of space.

It has been known for a long time (Crompton, 1973) that parasitic helminths do not occur uniformly or randomly throughout the alimentary tract of vertebrates. Every species shows a distinct preference for a particular region of the tract: numbers reach maximum density in one part of the tract and then decline in both anterior and posterior directions (Kennedy *et al.*, 1976). Such a distribution pattern directly parallels that of a free-living organism, which attains its maximum density in the area in which its resources are optimal and for which the density declines as resources become progressively suboptimal. In respect of helminths in the alimentary tract, it has gradually come to be accepted, especially subsequent to a paper by Holmes (1973), that the site of a species does correspond to its niche. It is possible, therefore, to quantify this niche by determining the mean (or median) position of a species along the tract, which is a measure of central location of numbers and the variance around that mean, which together with the range quantifies the niche of that species.

When several species are found in the alimentary tract of the same host species, it can be seen that their niches are distinct even though they may overlap (Fig. 7.5). In this example of the distribution of four acanthocephalan species along the alimentary tract of eels, the mean position of each species is distinct and differs significantly from that of every other species even though the overlap in distribution between some pairs of species is considerable (Table 7.8). The range occupied by each species is extensive and individuals can be found throughout almost the entire length of the alimentary tract. Put another way, each species occupies a different niche, although the niches of adjacent species overlap to differing extents (cf. Fig. 7.4). These data are illustrated at a component community level: it is important to appreciate that there is extensive variation in species distributions at infracommunity level and that in any individual host the patterns of distribution may differ considerably. For example, individuals of *Acanthocephalus lucii* may be spread throughout the range of the species. or they may be grouped at the anterior, median or posterior part of the range. In a similar manner, the niches of other species of acanthocephalans in fish, bird or mammalian hosts can be described and quantified (Kennedy, 1985).

When a species occurs on its own in the alimentary canal, its distribution corresponds to the fundamental niche of the species, i.e. the niche it occupies in the absence of any other species. This niche may expand as the density of the infection increases (Kennedy & Lord, 1982), but even on its own a species shows a preference for a particular section of the alimentary tract. The reasons for this are not clear. Price (1980) believes that this represents the normal adaptation of

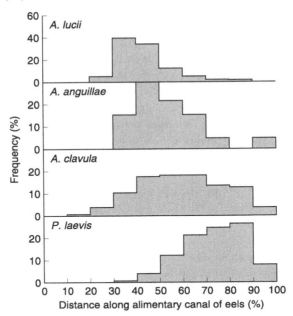

Fig. 7.5 The distribution of acanthocephalan species along the
alimentary canal of eels *Anguilla anguilla*. Data are from all fish
combined and so are appropriate to the component community level.
(From Kennedy, 1985b.)

a species to a particular set of resources, i.e. no species can adapt to all
possible niches. Rohde (1993, 1994) believes that restriction
of individuals to a particular niche will, in association with overdispersion, increase the probability of finding a sexual partner and
so of cross-fertilisation. Holmes (1973), however, argues that niche
restriction may represent selective site segregation by the species:
it arose as a means of avoiding interspecific competition in the past
and over evolutionary time it has evolved as a species characteristic.
This is the 'ghost of competition past' hypothesis: it provides a
satisfactory explanation, but is incapable of falsification (Simberloff,
1990). Which view one takes relates very much to the view that is
taken of interspecific competition: whether it occurs at all and if
so, whether it is of real importance in structuring helminth communities in the past or present or is just an occasional occurrence
of no ecological significance. Price (1980) and Rohde incline to this
latter view, on the basis that communities are seldom, if ever,

Table 7.8. *The position of acanthocephalans in the alimentary canal of eels* Anguilla anguilla

(a) Based on host population data

Species and site	No. of parasites	Mean position		95% CL	s^2	Range (%)
		%	SD			
Acanthocephalus clavula	488	60.5	17.6	1.56	309.7	19.2−100
Acanthocephalus lucii	652	43.2	17.5	1.33	304.5	7.2−82.2
Acanthocephalus anguillae	18	52.7	14.8	7.10	218.4	34.1−91.0
Pomphorhynchus laevis	171	73.1	13.3	1.99	176.9	39.0−100

(b) Based on individual host data

Species	Mean % position of population	SD and s^2 of mean % in individual eels around population mean		Range of individual	
		SD	s^2	Means	Variances
A. clavula	60.47	14.34	205.70	36.2−90.2	4.6−315.8
A. lucii	43.16	8.46	71.71	23.9−69.0	23.6−252.5
A. anguillae	52.70	14.77	218.40	34.1−91.0	all single
P. laevis	73.10	8.07	65.07	56.0−87.9	28.6−585.6

SD, standard deviation; CL, confidence limit; s^2, variance.
Source: From Kennedy (1985b).

saturated, that vacant niches are common and that there is no need to invoke competition when there is no evidence of resources being in short supply. By contrast, Holmes (1961, 1962a, 1973) and his co-workers (Bush & Holmes, 1986a,b; Stock & Holmes, 1988) argue that in many helminth communities, especially rich ones in aquatic birds, there is evidence that the communities are saturated with species, there are no vacant niches and clear examples of inter-specific competition have been demonstrated.

To some extent a synthesis of these views was proposed by Holmes & Price (1986), when they recognised that helminth communities could be arranged along a gradient. At one end are species poor communities, which are unsaturated and in which vacant niches exist and competition is unlikely: at the other extreme to these isolationist communities are the interactive communities which are species rich and packed and in which the realised niche of a species suggests that interactions between species are occurring in ecological time (Table 7.1). Indeed, intestinal helminths provide some of the main examples of communities in which interactions between species are primary processes determining the distribution and relative abundance of the species (Cornell & Lawton, 1992). This appears to be in part at least because it is possible to quantify the niche of each species and to see whether the realised niche differs from the fundamental niche of a species in the presence of another species. Such a functional response as a niche shift is generally accepted as evidence of competition occurring between species. So also are numerical responses, such as the regular absence of one species in the presence of another, i.e. competitive exclusion, or an inverse relationship between the density of one species and another in the same host or the adverse effects of one species upon growth and fecundity of another. Dobson (1985) has investigated the population dynamics of competition between parasite species using mathematical models, and has demonstrated that the greater the degree of overdispersion exhibited by each species the less will be the probability of their competing since the probability of heavy burdens of both species occurring in the same host individual will be low. He has also stressed that competition may be asymmetrical with only one species affecting the other(s), or it may be reciprocal. Both exploitation and interference competition may occur involving acanthocephalans. It is thus agreed by him and many other workers that interspecific competition can occur between helminths in the alimentary tract of vertebrates and that many of the best examples of this involve species of acanthocephalans.

7.5 INTERSPECIFIC INTERACTIONS

For a number of reasons, it might be predicted that interspecific interactions between species of acanthocephalans would be uncommon. Where, as is often the case, one species of acanthocephalan dominates a locality (Table 7.2), the probability of two species occurring in the same host individual must be very low. Moreover, the species of acanthocephalans exhibit resource partitioning in space (Fig. 7.6) such that no two of the British fish freshwater species

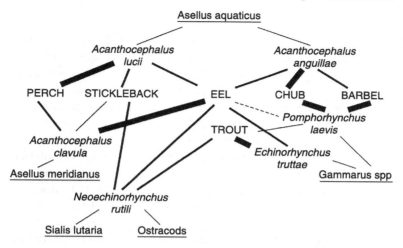

Fig. 7.6 A schematic diagram of the interrelationships between the
six species of acanthocephalan infecting British freshwater fishes (italics)
and their intermediate (underlined) and definitive (capitalised) hosts.
Thickness of lines joining the parasites to definitive hosts indicate
the relative importance of those hosts. Host–parasite relationships in
Ireland have been omitted for clarity. (From Lyndon & Kennedy, 2001,
with permission.)

share both preferred definitive and intermediate hosts (Lyndon &
Kennedy, 2001). Further, Dobson (1985) has demonstrated that the
probability of interspecific competition occurring between species
utilising the same host population is minimised as the degree of
overdispersion of each species increases, because then the probability of large infrapopulations of both species occurring in the
same host individual will decline. Finally, niche segregation will
minimise the extent to which any two species utilise the same
resources in the host. Nevertheless, acanthocephalan species may
co-exist in the same host individuals if they are forced to share a host
species, for example if their preferred host species is absent from
a region or locality, as happens for some species of acanthocephalan
in Ireland (Lyndon & Kennedy, 2001; Byrne et al., 2003). They may
also, of course, co-occur in individual hosts with species of other
helminth phyla, some of which, and particularly cestodes, are in the
same guild.

One of the earliest and best documented examples of interspecific competition involving an acanthocephalan comes from the

classic studies by Holmes (1961, 1962a,b) on experimental infections of *Moniliformis dubius* and the cestode *Hymenolepis diminuta*. Both species are in the same guild of absorbers and both species initially establish near the middle of the intestine in rats, and then migrate forwards to their preferred site at the anterior end of the tract. In a single species infection, *H. diminuta* has reached its preferred site by two weeks post infection and *M. dubius* by three or four weeks. Size and position of the cestode, but not of *M. dubius*, are affected by intraspecific competition (Holmes, 1961). The fundamental niches of the two species are almost identical (Fig. 7.7). If both species are given to a rat simultaneously, or if *H. diminuta* is administered before *M. dubius*, the cestode initially migrates to its anterior position, but as soon as *M. dubius* moves forward it moves posteriorly and by three to four weeks post infection it comes

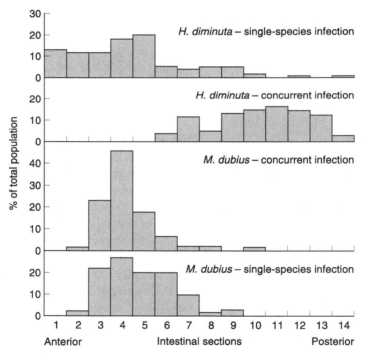

Fig. 7.7 Longitudinal distributions of the cestode *Hymenolepis diminuta* and the acanthocephalan *Moniliformis dubius* in the intestine of rats, showing the effects on niche selection of concurrent infections (data from Holmes, 1961). (From Whitfield, 1979, with permission, p. 226, copyright Hodder Education.)

to lie in the middle or posterior end of the intestine (Fig. 7.7). If *M. dubius* is already present in its preferred site in a rat and the rat is then given an infection of *H. diminuta*, the cestode remains in its initial position of establishment in the mid-gut and does not migrate anteriorly (Holmes, 1962a). In mixed species infections, therefore, the location of *M. dubius* is unchanged and it remains in its fundamental niche but *H. diminuta* shifts its niche. Its realised niche is less satisfactory, as evidenced by a decline in growth rate and size of the cestode, paralleling the effects of intraspecific competition, and the extent of the effect is related to the biomass of *M. dubius* present. A niche shift of this magnitude provides clear evidence of interspecific asymmetric interference competition, probably for space and access to nutrients such as carbohydrates, with the acanthocephalan emerging the winner. This competition only occurs in rats and there is no evidence that the two species compete in hamsters (Holmes, 1962b).

Interactions involving two species, *M. moniliformis* and the nematode *Nippostrongylus brasiliensis*, in the intestine of rats have been investigated in experiments by Holland (1984). Rats harbouring a 35-day-old infection of *M. moniliformis* were infected with *N. brasiliensis*; 10 days later significantly fewer nematodes had established and at maturity egg production of the nematode was reduced in comparison with single species infections. This period of 10 days is the normal time for the host's immune response to *N. brasiliensis* to start becoming effective. There was no change in weight or position of *M. moniliformis* in concurrent infections and the acanthocephalan was completely unaffected by the nematode. It is possible that the immune response to *M. moniliformis* (Andreassen, 1975a,b) is somehow involved in initiating non-reciprocal cross-immunity, or that the effect of the acanthocephalan on the intestine is inflammatory and renders the region unsuitable for other species. When three species, *M. moniliformis*, *N. brasiliensis* and *H. diminuta* are administered to rats, the interactions are more complex (Holland, 1987). Rats with 35 days post-infection doses of *M. moniliformis* and *H. diminuta* were infected with *N. brasiliensis*. The acanthocephalan and the cestode behaved as described by Holmes (1962a): *M. moniliformis* remained in its preferred anterior position with some weight loss, while *H. diminuta* exhibited stunting in growth and development in its less favoured position. There was also a significant reduction in the numbers of *N. brasiliensis* that could establish, which was again considered to be a possible example of non-reciprocal interference competition. Again, *M. moniliformis* emerged as the dominant species as neither its position nor density were affected by the other two species, yet it was able to interfere asymmetrically with both the cestode and the nematode in two or three species combinations.

Interspecific interactions between an acanthocephalan and a nematode have also been reported from a naturally occurring bird host. Dezfuli *et al.* (2002) investigated the relationship between the acanthocephalan *Southwellina hispida* and the nematode *Syncuaria squamata* in natural infections in cormorants, *Phalacrocorax carbo*. They found a strong negative correlation between the intensities of the two species across a sample of hosts (Fig. 7.8). Moreover, the size of the acanthocephalan became more variable in conjoint infections with the nematode, and vice versa. The effect on intensity appeared to be asymmetrical and non-reciprocal, whereas the interspecific density-dependent effect on growth was apparently reciprocal and symmetrical.

Many of the best examples of interspecific competition involving an acanthocephalan, however, come from studies on fish hosts. Grey & Hayunga (1980) studied the location of the acanthocephalan *Pomphorhynchus bulbocolli* and the cestode *Glaridacris laruei* in naturally

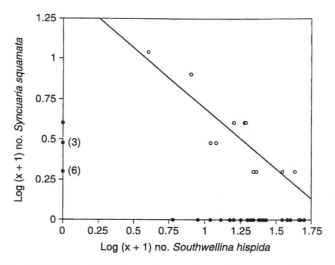

Fig. 7.8 Relationship between the numbers of the nematode *Syncuaria squamata* and the acanthocephalan *Southwellina hispida* in 11 cormorants that harboured both species (hollow circles). Data for 40 birds harbouring only one species are also shown (solid circles). Numbers in parentheses indicate stacked symbols. (From Dezfuli *et al.*, 2002.)

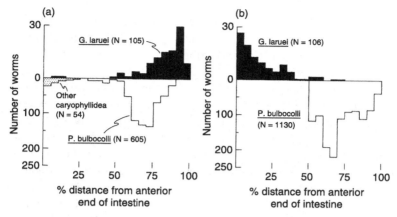

Fig. 7.9 Distribution of intestinal helminths in fish. (a) In eight fish in which the density of *Pomphorhynchus bulbocolli* was relatively low (mean = 76, max. = 144) and the cestode *Glaridacris laruei* was found in its usual habitat, and (b) in four fish in which *P. bulbocolli* was more abundant (mean = 283, max. = 648) and occupied the entire posterior of the gut. This displaced *G. laruei* from its usual posterior position to the anterior of the gut. (From Grey & Hayunga, 1980, with permission.)

occurring single species and conjoint infections in suckers *Catostomus commersoni*. The preferred site of the cestode is in the posterior half of the intestine, as is the preferred site of the acanthocephalan. When infection intensities of *P. bulbocolli* are low, both species co-exist in their same preferred region of the intestine, but when the intensity of the acanthocephalan is high, the cestode is now found in the anterior region of the intestine (Fig. 7.9). This niche shift is asymmetrical and non-reciprocal, as the acanthocephalan appears to be completely unaffected by the cestode. Here the specialist acanthocephalan is able to exclude the more generalist cestode from its preferred site. A rather similar niche shift has been reported by Chappell (1969) in naturally occurring concurrent infections of three-spined sticklebacks *Gasterosteus aculeatus* with the acanthocephalan *Neoechinorhynchus rutili* and the cestode *Proteocephalus filicollis*. In single species infections both species of parasite are widely dispersed throughout the intestine, although the cestode shows a preference for the more anterior region and the acanthocephalan for the more posterior. In concurrent infections the overlap in distribution between the two species is significantly reduced, as *N. rutili* attaches in a more posterior location and *P. filicollis* in a more anterior one. This niche shift by the two species was interpreted as being an example of reciprocal and symmetric interspecific competition, with each species retreating to its preferred niche in the presence of the other.

The distributions of the acanthocephalans *Acanthocephalus anguillae* and *Pomphorhynchus laevis* in the British Isles are discrete and discontinuous (p. 42), and Kennedy *et al.* (1989) suggested that the two species might compete as they showed a preference for the same definitive host, the chub *Leuciscus cephalus*, and had similar effects upon the host intestine. Bates & Kennedy (1990) investigated this possibility further in laboratory infections of rainbow trout *Oncorhynchus mykiss*: a host in which both species could attain full sexual maturity but which, being an introduced species, was not the natural host for either. At low levels of infection, the distribution of the two species in the intestine showed a 44% overlap and there was no evidence of competition between them. At higher levels of infection, establishment of each species was unaffected by the other, but the survivorship (Table 7.9) and range of intestine occupied by *A. anguillae* was progressively reduced as intensities of *P. laevis* increased (Fig. 7.10). The position of *P. laevis* was unaffected in concurrent infections so the interaction was asymmetrical. The findings fit the criteria suggested by Dobson (1985) for the asymmetrical effect of interference competition as it appears that *P. laevis* is able to create

Table 7.9. *Establishment and survival of* Acanthocephalus anguillae *from experimental infections of* Oncorhynchus mykiss *with this species and* Pomphorhynchus laevis

	Recovery of *A. anguillae*					
	7 days p.i.		56 days p.i.		112 days p.i.	
	Mean no recovered ($\pm$ SE)	% established	Mean no recovered ($\pm$ SE)	% surviving from 7 dpi	Mean no recovered ($\pm$ SE)	% surviving from 7 dpi
15 *A. anguillae* + 45 *P. laevis*	5.4 (2.15)	36.0	3.3 (0.63)	61.1	0.0	0.0
30 *A. anguillae* + 30 *P. laevis*	15.8 (3.47)	52.7	5.9 (0.57)	31.6	2.7 (1.25)	18.1
45 *A. anguillae* + 15 *P. laevis*	16.8 (3.47)	37.3	13.2 (5.80)	78.6	6.5 (4.27)	38.7
60 *A. anguillae*	16.8 (4.03)	28.0	9.2 (2.06)	55.0	8.2 (2.46)	48.8

Source: From Bates & Kennedy (1990).

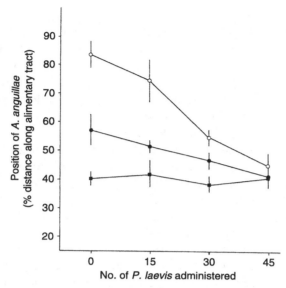

Fig. 7.10 The mean position of *Acanthocephalus anguillae* 56 days post-infection in high density infections in conjoint experimental infections in *Oncorhynchus mykiss*. Solid circles, mean position; open circles, posterior; squares, anterior. (From Bates & Kennedy, 1990.)

an exclusion zone around itself, similar to that reported for *Moniliformis dubius* by Burlingame & Chandler (1941). Interestingly, and inexplicably, paralleling the results of Holmes (1962b), interspecific competition between these same two species could not be demonstrated in other species of fish, including the preferred host chub (Bates & Kennedy, 1991b).

Other examples strongly suggestive of interspecific competition and involving functional or numerical responses have been reported in association with acanthocephalans in fish. Kennedy (1992) suggested that *Acanthocephalus lucii* and *A. anguillae* could compete in eels *Anguilla anguilla* under some circumstances. Each species could occupy a large region of the intestine, and niche overlap was high even though *A. lucii* had a more anterior mean position. In concurrent infections at co-equal densities, the mean position of each species was unchanged but the variance around this of both species declined so that niche overlap was reduced. This suggested that exploitation competition was occurring. At high densities of *A. lucii* and low densities of *A. anguillae* reproduction of this latter species declined as the numbers of *A. lucii* in the more anterior position blocked the movement of *A. anguillae* down the intestine to its preferred site and so the probability of individuals mating declined. This was not believed to be a competitive interaction in any way. However, the combination of the two interactions could explain the dominance of *A. lucii* in localities in Ireland in which the two species co-occurred (Kennedy & Moriarty, 1987). No interaction between the two species has been reported in England (Norton *et al.*, 2004a,b).

Byrne *et al.* (2003) have provided evidence for negative interactions between *Acanthocephalus clavula* and *Pomphorhynchus laevis* in natural populations of brown trout *Salmo trutta* in Ireland. Outside Ireland, co-infections of fish species with these two species of acanthocephalan are very uncommon. They tested the frequencies of co-occurrences against those generated by a null model, but found no significant differences between the observed and predicted values. However, they found a significant negative association between the numbers of each of the two species in concurrent infections; the niche width of *P. laevis* decreased markedly in high intensity concurrent infections with *A. clavula*, resulting in a decline in niche overlap. Following Lyndon & Kennedy (2001) they also considered that acanthocephalan interactions are likely to be observed more frequently in Ireland because a number of possible host species are absent from the country and so acanthocephalan species are forced to co-exist in hosts more often than in other parts of Europe, and they often dominate the helminth communities there.

Dezfuli *et al.* (2001a) also studied species co-occurrences and interspecific associations in intensity in helminth communities in *S. trutta* in Italy. The communities in the three localities surveyed were dominated by acanthocephalan species. Frequencies of co-occurrences did not differ from the predictions of a null model, but there were a number of negative correlations in tests of pairwise associations. However, none of these significant associations was found in more than one host population, suggesting that the correlations were very condition dependent. In apparently similar localities, the same species combinations showed different associations. The most significant associations were found between *Acanthocephalus anguillae* and *Echinorhynchus truttae*, between *A. anguillae* and *Pomphorhynchus laevis* and between *P. laevis* and *E. truttae* (Table 7.10). These were also the three commonest species of acanthocephalans. All three species also showed negative associations with the cestode *Cyathocephalus truncatus*. They considered that these findings were consistent with interspecific

Table 7.10. *Matrix of pairwise associations (Spearman's rank correlation coefficient) between the intensity of infections of four helminth species in trout* Salmo trutta *from three streams in northern Italy*

Coefficients are shown above the diagonal and sample sizes (numbers of fish harbouring at least one of the species) below it.

	P. laevis	A. anguillae	E. truttae	C. truncatus
Tergola				
Pomphorhynchus laevis	—	−0.360	−0.119	−0.118
Acanthocephalus anguillae	23	—	−0.367	−0.192
Echinorhynchus truttae	17	18	—	−0.230
Cyathocephalus truncatus	33	34	32	—
San Giorgio				
P. laevis	—	0.083	−0.100	−0.222*
A. anguillae	53	—	−0.530***	−0.150
E. truttae	53	39	—	−0.276*
C. truncatus	56	36	42	—
Grimana Nuova				
P. laevis	—	−0.803***	−0.676*	−0.535**
A. anguillae	8	—	—	−0.472**
E. truttae	7	2	—	−0.383*
C. truncatus	23	20	20	—

Source: From Dezfuli *et al.* (2001a), with permission.

competition between the species pairs, but without evidence of replicability it is almost impossible to form any conclusions about the consistent role of competition between species in the field.

The only report of interspecific interactions in tropical localities comes from the study of helminth communities in natural populations of *Cichlasoma synspillum* in Mexico by Vidal-Martinez & Kennedy (2000b). They analysed positional data on a community of 18 helminth species in this host in one locality. There was only a single species of acanthocephalan present, *Neoechinorhynchus golvani*, but this species was involved in interactions with the digenean *Crassicutis cichlastomae*, and the nematodes *Spirocamallanus rebeccae* and *Rallietnema kritscheri*. All four species were host specialists. The interactions were expressed as negative correlations, positional shifts and reductions in realised niches in the presence of the acanthocephalan. It appeared that the acanthocephalan produced unsuitable habitats in the intestine for the other three species, i.e. an exclusion zone, but its own position and intensity were unaffected by any other species. These asymmetrical interactions between unrelated species strongly suggested interspecific competition, with the acanthocephalan as usual being the winner.

Nothing at all is known about interspecific interactions between acanthocephalan species in their paratenic hosts, and relatively little about interactions in intermediate hosts. Larval stages in intermediate hosts often occur at low (<1.0% or 0.1%) levels of prevalence (Denny, 1969; Amin, 1978) and so the probability of finding a concurrent infection is normally very low. Awachie (1967) reported that only three of several thousand gammarids harboured conjoint infections of *Echinorhynchus truttae* and *Polymorphus minutus* in a small stream. Van Maren (1979b) reported the co-occurrence of *Polymorphus minutus* with *Pomphorhynchus laevis*, *P. minutus* with *M. truttae* and *P. laevis* with *M. truttae* in *Gammarus fossarum* in the River Rhone without any apparent effects of one species on another. Dezfuli *et al.* (2000) also studied the associations between acanthocephalans in an intermediate host *Echinogammarus stammeri* in a small stream: prevalence levels of *Pomphorhynchus laevis* were 18.5%, of *Acanthocephalus clavula* 1.3% and of *Polymorphus minutus* 0.9%. The first two species shared the same definitive host and there was a positive association between them, so any manipulation of the intermediate host behaviour would benefit both and this could lead to the association being transferred to the definitive host. However, the levels of conjoint infections were so low that this might be of little or no significance. Reduction in cyst size of *P. laevis* was observed in co-infections with *A. clavula* (Dezfuli *et al.*, 2001b), but there was no indication that this had any deleterious effect on the parasite. Barger & Nickol (1999) have examined the effects of experimental co-infection of *Hyalella azteca* with larvae of *Pomphorhynchus bulbocolli* and *Leptorhynchoides thecatus*. A significantly smaller proportion of *L. thecatus* reached the cystacanth stage in conjoint infections. However, this may be of little ecological significance as habitat separation by eggs (Barger & Nickol, 1998) will tend to reduce the frequency of occurrence of concurrent infections in nature.

Overall, no clear pattern of the role of acanthocephalans in helminth communities emerges. This is partly because there is very little firm information on the great majority of species and conclusions are inevitably based on a small number of freshwater species in temperate fish. These may not be representative. Helminth communities in fish are generally species poor and exhibit low diversity, and many of them appear to be dominated by an acanthocephalan species. This contrasts with the situation in many birds and mammals, where helminth communities are species rich and packed and diversity is much higher: here acanthocephalans are often satellite species that may contribute to species richness but contribute little or nothing to community structure. It is clear that acanthocephalans do behave as predicted by niche theory in some situations. They interact with other species and they can exhibit both exploitation and interference competition in the intestines of vertebrates, but the wider significance of this is still unclear and there is disagreement on whether interspecific competition plays an important structuring role in helminth communities in nature. As Dezfuli *et al.* (2001a) pointed out, the occurrence of interspecific competition may be very dependent on local conditions, and what happens in one locality may not happen in another. Nevertheless, it has been shown convincingly that interspecific competition involving acanthocephalans can occur in species poor, isolationist communities and that acanthocephalans can compete with other species of acanthocephalan and with cestodes, digeneans and nematodes in all classes of vertebrate host. The impression that is conveyed by the evidence is that their zone of exclusion means just that: acanthocephalans do not suffer other species of their own or other kinds to live near them in their fundamental niche.

8

Introductions and extinctions

8.1 DISPERSAL

Introductions, colonisations of new habitats and extinctions are all closely related to the dispersal ability of a species. It is seldom appreciated that a colonisation event and an extinction event may be indistinguishable at the moment they are viewed: from the perspective of a parasite, the same strategy is involved both in establishing a population from a colonising propagule and in retrieving a population from the point of extinction. Introductions and extinctions are in effect the two sides of the same coin and the success or failure of each process is closely linked to the dispersal ability of the parasite.

In his review of parasite dispersal, Kennedy (1976) concluded that the so-called dispersal stages of parasites were not particularly effective in achieving dissemination in space or time. The duration and timing of the reproductive period and the release of eggs of a parasite, especially if reproduction exhibits a seasonal cycle, is more closely related to the opportunities for, and probability of, infecting the next host in the life cycle than to anything else. If the intermediate host of an acanthocephalan itself exhibits a seasonal cycle in reproduction such that the new generation appears at a particular time of year, then parasite reproduction will normally be synchronised with this. The activities of any short-lived, active, free-living stages of a parasite are also far more likely to be related to the probability of locating and infecting the next host than to disseminating the parasite in space or time, as they cover only very short distances in their lifespan. Parasites with inactive free-living stages, such as the eggs of acanthocephalans, are unable to achieve much dissemination in space by their free-living stages, although they can achieve some dispersal in time. However, most dissemination in space and time is achieved by the movement and longevity of their hosts, both intermediate and definitive. Acanthors can generally survive for only a few weeks (exceptionally for two to three years) (Table 2.2), but cystacanths can generally survive as long as their intermediate hosts. Whereas most invertebrates used as intermediate hosts can survive for up to or over a year, many definitive hosts survive for far longer periods than this, and the inclusion of paratenic hosts in the life cycle will further extend an acanthocephalan's lifespan. The distances achieved by the natural movements of vertebrates far exceed the distances that eggs can be carried by wind action or by aquatic drift. Reliance on host mobility for dissemination removes the need for the acanthocephalans to disperse in space or time and is far more effective in distributing eggs if the parasites themselves are iteroparous.

This is exemplified by the distribution of *Polymorphus minutus* in a small stream (Table 8.1). The source of the infection was a population of ducks, which were located in a pen at point 0 m. Infection levels of *Gammarus pulex* fell off rapidly upstream, but the infection could still be detected in gammarids up to 5 km downstream, albeit at a low level (similar to that 50 m upstream). The downstream spread is likely to result from downstream drift of acanthors, but especially from selective downstream drift of infected *G. pulex* (Fig. 5.7). Van Maren (1979a,b) similarly found that at one site on a tributary of the River Rhone where populations of *Gammarus fossarum* were very dense and ducks abundant, prevalence of *Polymorphus minutus* was as high as 48%, but 100 m downstream only 2% were infected, and below this region no infected gammarids were found. However, these distances are negligible compared with the distances that can

Table 8.1. *Infection of* Gammarus pulex *with* Polymorphus minutus *above and below a duck pen on a small stream*

	Distance in metres from duck pen						
	Above pen		Below pen				
	300	50	10	250	800	2000	5000
% infected	3.0	15.1	70.9	62.9	66.2	19.9	15.6
Mean intensity	1.1	1.8	3.3	2.3	2.0	1.3	1.1

Source: Data from Hynes & Nicholas (1963).

be covered by the ducks. In a natural, as opposed to a farm, situation where ducks undertake seasonal migrations, the acanthocephalans they harbour can be transported across continents, whereas movement of *G. pulex* cannot take them outside the river catchment as the amphipod is unable to survive in estuarine conditions.

Dispersal will also be assisted by anthropochore movements of host population, as, for example, for restocking of fisheries or recolonisation of habitats, and the rapid transit possible now as a result of air-freighting enables hosts with their parasites to be moved easily and rapidly around the world. Transoceanic and transcontinental journeys by land or by sea taking weeks or months have now shrunk to days or hours, minimising the loss rate of parasites over the time of the journeys. Eels and their parasites, for example, can be, and are now, regularly flown live from Australia to Hong Kong and Japan directly in a matter of hours. Live eels from New Zealand have been identified in a market in London and have been found to contain a number of New Zealand species of helminth, though none has yet been found to have established itself in Britain (P. J. Whitfield, pers. comm.).

Acanthocephalans therefore have considerable dispersal potential as a consequence of movements of their host(s), but as has been emphasised previously (pp. 35, 132–3, 138) there is extensive local variation in the presence and abundance of any species (Tables 3.6, 3.7). In this respect, the acanthocephalans present a paradox. Seven species are particularly widespread and common in freshwater fish throughout the British Isles and most of north-western continental Europe. *Metechinorhynchus salmonis* is present throughout much of northern Europe but it is now considered never to have been present in the British Isles (Chubb, 2004) and it is absent from there and from southern Europe because of the restricted distribution of its intermediate host, the glacial relict *Pontoporeia affinis*. Despite this widespread regional distribution of the seven species, the local occurrence of each is very patchy and adjacent localities may harbour very dissimilar acanthocephalan communities. Esch *et al.* (1988) have drawn attention to distinct colonisation strategies exemplified by parasites as being a possible explanation for the apparently stochastic nature of freshwater fish helminth communities and the erratic and unpredictable local occurrence and distribution of many species. They distinguish between autogenic species which mature in fish and allogenic species which mature in vertebrates other than fish and so have a wider dispersal ability. Autogenic species are restricted by land or sea barriers to particular habitats, but allogenic species have greater vagility and are able to cross such barriers. They further demonstrated that hosts with helminth communities dominated by autogenic species are likely to show lower levels of community similarity than those dominated by allogenic species (Fig. 8.1).

Many acanthocephalans, and especially those maturing in freshwater fishes, are autogenic. However, some species have overcome this limitation by their ability to utilise one or more species of a migratory fish as a preferred or suitable host in their life cycle, allowing transfer of an acanthocephalan species from one catchment to another and so colonisation of new catchments (Lyndon & Kennedy, 2001). Each of the six species infecting British fish can use a colonisation host in this manner (Fig. 7.6). However, even with this ability, acanthocephalans are still restricted in their ability to disperse. Following an environmental disaster in a small and isolated

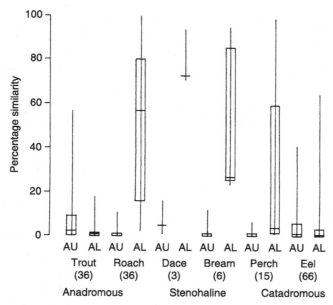

Fig. 8.1 Summary of matrices of percentage similarity index comparisons of helminth fauna of each species of fish between localities. Horizontal lines indicate median, boxes the 25% and 75% quartiles, and vertical lines the range. Numbers refer to the number of paired comparisons. AU, autogenic; AL, allogenic. (From Esch et al., 1988.)

Table 8.2. *Decline of* Acanthocephalus clavula *in* Perca fluviatilis *in* Slapton Ley

	Year and month				
	1979	1979	1980	1981	1984
	May	Oct.	May	Aug.	May
No. of fish	30	38	41	22	0
Prevalence (%)	26.6	42.1	7.3	0	0
Mean abundance	0.43	1.89	0.15	0	0
Max. intensity	6	15	3	0	0

lake that caused fish populations to collapse below the level at which helminth parasite populations were able to transmit to them (Kennedy, 1998), some allogenic species and most autogenic species (Table 8.2) became locally extinct. All the allogenic species, which used wild birds as definitive hosts, were able to recolonise the lake naturally, confirming the superior dispersal ability conferred by this strategy, but although the fish populations recovered without any restocking neither *A. clavula* nor any of the other autogenic species reappeared in the lake (Table 8.3).

Dispersal alone may not be sufficient in itself to ensure colonisation or to promote similarity between acanthocephalan communities. The River Exe is a large river that is joined by several tributaries along its length. It should therefore be possible for the acanthocephalan species in one river to disperse to another either by drift of the intermediate hosts or movements of fish within the catchment. In reality there is considerable local variation in the presence and abundance of acanthocephalan species in the different tributary rivers (Table 8.4). *Echinorhynchus truttae* is only found in two tributaries and then at low levels, whereas *Acanthocephalus lucii* is only found in one, and only one river is dominated by *Pomphorhynchus laevis*. It is likely that much of this variation relates to the distribution of hosts (Kennedy, 2001), as *Gammarus pulex* is common only in the River Culm and brown trout are less common in the lower reaches of the Exe; but, whatever the reasons, it shows clearly that dispersal ability does not necessarily lead to the establishment of a population. A parallel to this situation has been reported for the rivers Severn and Lugg, a tributary of the River Wye. The dominant acanthocephalan in the Severn is *Pomphorhynchus laevis* (Brown, 1986, 1989), but even though the Wye joins the Severn, *P. laevis* was very

Table 8.3. *Changes in the composition of the helminth parasite fauna of fish in Slapton Ley before and after the fish population crash in the winter of 1984/1985*

| | Year | | | | | | | | | |
	<1973	1973	1977	1984//	1990	1991	1992	1993	1994	1996
Autogenic										
Tetraonchus monenteron	P	P	P	P	P	P	P	P	P	P
Caryophyllaeus laticeps	P	P	P	A	A	A	A	A	A	A
Caryophyllaeides fennica	P	P	P	P	A	A	A	A	A	A
Acanthocephalus lucii	P	P	A	A	A	A	A	A	A	A
A. clavula	P	P	P	A	A	A	A	A	A	A
Allogenic										
Ligula intestinalis	A	P	P	P	P	P	P	P	P	P
Diplostomum spathaceum	P	P	P	P	P	P	P	P	P	P
D. gasterosti	P	P	P	P	A	P	A	A	P	A
Tylodelphys clavata	A	P	P	P	P	P	P	P	P	P
T. podicipina	A	A	P	P	A	P	A	A	P	P

Source: Data from Kennedy (1998).

Table 8.4. *The relative abundance of each acanthocephalan species as a proportion (p_i) of the total number of all acanthocephalans of all species in eels* Anguilla anguilla *in the River Exe catchment and in the adjacent River Otter*

| | Rivers | | | | | | |
	Creedy	Exe B	Exe S	Alphin	Clyst	Culm	Otter
Neoechinorhynchus rutili	0.98	0.75	1.0	0	0.25	0	0
Echinorhynchus truttae	0.02	0	0	0	0	0.01	0
Acanthocephalus lucii	0	0.25	0	0	0	0	0
Acanthocephalus clavula	0	0	0	0	0.66	0.04	0
Pomphorhynchus laevis	0	0	0	0	0.09	0.95	1.0

Source: From Kennedy (2001).

uncommon in chub *Leuciscus cephalus* and restricted in its distribution in the River Lugg. However, 20 years later, its presence at the same site in the River Lugg was confirmed and prevalence levels in *Gammarus pulex* were 10.6%, yet the parasite was still not found in any sampling sites along the River Wye (Kennedy *et al.*, 1989).

A similar situation was apparent in the River Rhone, where Van Maren (1979a,b) found sites to be dominated by *Pomphorhynchus laevis* where the flow was fast and *G. fossarum* abundant, but in backwaters and where the flow was less, *Asellus* species were more common and so accordingly was *Acanthocephalus anguillae*. In one tributary where burbot *Lota lota* were present, so also was *Asellus meridianus* and the dominant acanthocephalan was accordingly *A. clavula*. In yet another tributary where grayling *Thymallus thymallus* was very common, *Metechinorhynchus truttae* rather than *P. laevis* was the dominant acanthocephalan. In the River Danube and most of its tributaries *Pomphorhynchus laevis* is the dominant acanthocephalan (Moravec *et al.*, 1997; Moravec & Scholz, 1991), but it is absent from the Jihlava river where *Neoechinorhyncus rutili* dominates even in barbel *Barbus barbus*: it is suggested that this may reflect a lower availability of the intermediate host in this locality.

Kennedy *et al.* (1991) and Hartvigsen & Kennedy (1993) investigated the importance of fish stocking in the dissemination of parasites throughout a group of reservoirs of rather similar age and chemical composition fairly close to each other in the same region of south-west England. They predicted that the anthropochore movements of brown trout *Salmo trutta* and rainbow trout *Oncorhynchus mykiss* between some neighbouring reservoirs for stocking and management purposes would similarly disseminate their parasites. This should result in a high degree of similarity between the parasite communities of fish in the water bodies that have been linked by such transfers, and a lower degree in the reservoirs which have received no recent stocking. What they found was that there was no significant difference in mean similarity between the helminth faunas of trout in the reservoirs linked by stocking and those not so linked. The acanthocephalan species, as might be expected, showed considerable local variation in composition and abundance of the two species present (Table 8.5), but this pattern was not consistent with their having been disseminated by fish stockings. In fact, the two reservoirs in which either of the two species was common fell into quite discrete cluster groups.

In this context the distribution of *P. laevis* in Britain is relevant. It is believed that the species was originally post-glacially confined to the River Thames (Kennedy *et al.*, 1989), and was dispersed to the River Severn and Hampshire's River Avon when these rivers were stocked with barbel *Barbus barbus* from a tributary of the Thames. However, although barbel were stocked into the Bristol Avon,

Table 8.5. *Comparison of the acanthocephalan fauna of brown trout* Salmo trutta *in a group of reservoirs in south-west England of which some have been subjected to recent fish stocking and others remained unstocked*

	Echinorhynchus truttae		Neoechinorhynchus rutili	
	%	Abundance (SD)	%	Abundance (SD)
Unstocked				
Avon Dam	0	0	0	0
Colliford	0	0	0	0
Meldon	0	0	0	0
Venford	8.3	0.8 (0.6)	0	0
Stocked				
Argal	0	0	0	0
Burrator	0	0	0	0
Crowdy	20.9	0.3 (0.9)	0	0
Siblyback	0	0	93.8	133.8 (199)
Stithians	0	0	7.7	0.1 (0.5)
Upper Tamar	0	0	0	0

Source: Reprinted from Kennedy *et al.* (1991), with permission from Elsevier.

the parasite did not establish there, possibly because *Gammarus pulex* occurred there only at low levels. Clearly, factors other than dispersal, including chance, are also essential to successful colonisation.

8.2 INTRODUCTIONS AND COLONISATION

The important issue in relation to introductions and colonisations is the extent to which the dispersal potential of an acanthocephalan species is actually realised. Introductions do not of necessity lead to successful colonisations and invasions are commoner than establishments (Kennedy, 1993c, 1994). MacArthur & Wilson (1967) have identified the desirable attributes of a good coloniser as being good dispersal ability, a high reproductive potential and hermaphroditism and/or asexual reproduction. The first attribute enables species to cross barriers to dispersal more easily, the second to build up a new population rapidly after introduction to a new locality and the third to found a new population with a small initial propagule. The importance of population size cannot be underestimated as the smaller the population size of an organism, the greater is the probability of stochastic extinction (MacArthur & Wilson, 1967). Since most colonising propagules are small, the danger of such extinction is inevitably very high. Survival of an introduced species is only possible if and when it reaches a threshold population size at which R_0 equals or exceeds 1 (Anderson & May, 1979; Anderson, 1982). A small propagule will also incorporate only a proportion of the gene pool of the species, and this founder effect may ultimately lead to genetic differences between populations and eventually to the formation of strains. However, if a heavily infected host individual is a part of the invading propagule, it may harbour a large proportion of the source host population (one duck has been reported as harbouring 90% of the individuals of *Polymorphus* sp. in a duck population).

A fourth desirable attribute is wide host specificity (Kennedy, 1994). A generalist species able to utilise a wide variety of definitive and intermediate host species is more likely to be a successful colonist than a species with narrow specificity. Colonisation ability can also be predicted to be inversely proportional to the complexity of a parasite's life cycle. Each host species must not only be present in a locality, but must be present at a density that permits it to survive and the parasite to transmit successfuly to it: the more hosts that are required, the more demanding these conditions become. Parasites with direct life cycles should therefore be more likely to be successful invaders. The obstacles to a successful colonisation can be envisaged as a series of barriers that have to be overcome (Fig. 8.2). The first barrier is the physico-chemical conditions. The second is the host species present in the new habitat, which must contain suitable species and at the requisite density, i.e. above the threshold, to serve as definitive and intermediate hosts, and the food web must be such as to permit transmission. The third barrier, that of parasite factors, includes the specificity of the parasite and its ability to infect hosts other than its preferred species (Kennedy, 1994). The greater the similarity of the new landscape to the source one, the greater is the probability of a successful colonisation (Fig. 8.2A): if the landscapes are congruent, the only barrier to a successful colonisation may be a stochastic one. Chance factors, such as timing, are important in determining the likelihood of an introduction succeeding, as the barriers may not be open all through the year. For example, if an acanthocephalan can only infect juveniles of its preferred arthropod species, and this species shows seasonality in reproduction, the acanthocephalan that arrives in winter may find no suitable intermediate hosts and the barrier closed (Fig. 8.2Bw) whereas if it arrived

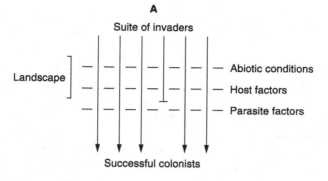

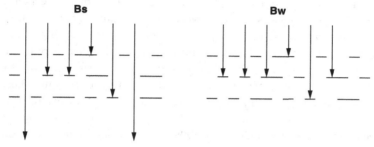

Fig. 8.2 The influence of barriers to successful colonisation and the concepts of a compatible landscape and transmission windows. A population of an introduced species of fish will harbour only a part of the parasite community in its source locality. Introduced into locality A, the parasites encounter similar abiotic and biotic conditions. The landscape is almost identical and so most species colonise. If the same suite is introduced into locality B in summer, Bs, parasites encounter barriers of unsuitable temperature, unavailable hosts etc. The landscape is only partly compatible, the transmission window is narrower and so only a few species colonise successfully. An introduction to the same locality in winter (Bw) may mean that no species will succeed, as transmission windows have now closed and the landscape is temporarily incompatible. (Based on an idea in Holmes, 1987; from Kennedy, 1994, with permission of the author.)

in spring or summer the barrier may be open (Fig. 8.2Bs). Even then, a shortage of time when infection is possible may prevent establishment. It may simply be a matter of chance when the parasite arrives in a new locality and whether the invading propagule is of the right size to succeed, but this stochastic element is seldom fully appreciated. There are more introductions than successful establishments (Kennedy, 1993c, 1994), and many acanthocephalans with the potential to colonise a new locality may reach it by chance at the wrong time when the transmission window is too narrow or is actually closed.

Acanthocephalans have, as seen above, good dispersal ability and are believed to be *r*-selected (Esch *et al.*, 1977), but they are not hermaphrodite and they are incapable of asexual reproduction. Moreover, they have indirect life cycles and so require the presence of two host species in the new potential habitat. By contrast, many species show wide specificity to their definitive hosts (Chapter 4) and so can infect a range of species other than their preferred hosts. A complex life cycle also means that a parasite is not at the mercy of a single host species for its survival: if, for example, the definitive host population suffers a temporary decline, the parasite may survive in the intermediate host until the definitive host population recovers. Even if all of the preferred hosts are absent from a new and strange land, a parasite may be able to utilise a related host instead (Kennedy & Bush, 1994): *Pomphorhynchus laevis*, for example, is able to utilise the introduced North American species *Oncorhynchus mykiss* in Europe (Kennedy *et al.*, 1978), as is *Echinorhynchus truttae* on the island of Jersey (Kennedy *et al.*, 1986b). However, it is impossible to generalise about this as the related species *P. patagonicus* prefers the endemic fish species of Patagonia as hosts and is not able to reproduce in *O. mykiss* even though it can infect it (Trejo, 1994; Ubeda *et al.*, 1994). Some species show great specificity to their intermediate host, to the extent of being monospecific. This is the case with *Acanthocephalus clavula* which uses *Asellus meridianus* in the British Isles and *A. lucii* which uses *A. aquaticus* and *Pomphorhynchus laevis* which uses *Gammarus pulex*. In these three examples the intermediate host can be abundant in a range of habitats and is often a key species in the food web there, thus facilitating transmission (Marcogliese, 2002). These same species are also often able to change their intermediate host in different regions, for example *A. clavula* uses *Echinogammarus pungens* in northern Italy (Dezfuli *et al.*, 1991a,b). Similarly, *P. laevis* uses *E. stammeri* in northern Italy (Dezfuli *et al.*, 1999) and *G. duebeni* in Ireland (Fitzgerald & Mulcahy, 1983). The chances of a successful establishment may also depend upon the identity of the potential host. These abilities should serve to minimise the potential disadvantages of a two-host life cycle.

Other considerations may also affect the success or failure of an introduction. The introduced acanthocephalan has to be capable of invading the helminth community of the existing host in the new locality. It will clearly be easier to invade a parasite community if the recipient community is species poor and isolationist in character, which is the case for fish and amphibian hosts and for some birds and mammals. It will also be easier to invade if vacant niches exist in the recipient community, as Price (1984, 1986) believes is generally the case. If vacant niches do not exist or/and the recipient communities are interactive, the invading parasite may have to outcompete a native species. Here again, generalist acanthocephalans stand a good chance of succeeding as they are often successful competitors against other species of helminths (pp. 148–51).

The colonisation successes of some acanthocephalan species are spectacular. The Hawaiian Islands are one of the most isolated archipelagos in the world as far as freshwater fish are concerned yet they are home to five native species of fish and four species introduced by humans (Font, 1997, 1998). The native fish species harbour three helminth species introduced by birds and four by marine fish. Among these introduced helminths is one species of acanthocephalan, *Southwellina hispida*, on Hawaii and Oahu. The larva of this species is found in both native and introduced fish species, which probably serve as paratenic hosts since the larvae are found in the body cavity. The adults were probably introduced by piscivorous birds, and probably by the black-crowned night heron *Nycticorax nycticorax* which has itself recently colonised the Island of Hawaii. By contrast, the recently introduced species of poeciliid fish, introduced by humans to assist in mosquito control, have failed to introduce any helminth parasites to the islands, including *Octospinifer chandleri* which commonly infects them in their source localities in North America. This failure may well be due to the absence of suitable intermediate hosts on the islands, or may just be a stochastic failure. Poeciliids have also been introduced into Australia, but have also failed to introduce any acanthocephalan species with them (Dove, 2000). However, the acanthocephalan *Plagiorhynchus cylindraceus* is believed to be an introduction into Australia, initially with introduced birds. Its intermediate hosts *Porcellio scaber* or *Armadillidium vulgare* have also been introduced into Australia by humans, and so after initially cycling through the introduced species it is believed to have spread to native species of Australian birds and thence even to marsupials, having now been found in insectivorous bandicoots (Smales, 1988; Beveridge & Spratt, 1996). In a similar manner, *Onicola pomatostomi* is believed to have been introduced to Australia by birds, which serve as paratenic hosts for the species in Malaysia and Borneo where the definitive hosts are leopard cats: in Australia the parasite uses dingos and feral cats as definitive hosts (Schmidt, 1983).

It is probable that successful colonisations are fairly regular events, but are seldom recognised as such. To identify them properly requires a time series of observations, often over a long period, if they are to be distinguished from short-lived, unsuccessful accidental occurrences. For example, *Pomphorhynchus laevis* has been reported to have a restricted distribution in England, where it is mainly found in only three river systems (Fig. 3.3a) having been introduced from one into the other two by humans (Kennedy *et al.*, 1989). In Ireland, it was widespread but was apparently restricted to the south of the country. However, it has very recently appeared in the north of Ireland, from *Oncorhynchus mykiss* raised in a fish farm and from *Anguilla anguilla* from the River Erne catchment (Evans *et al.*, 2001). The records appear to be independent of each other and while its appearance in rainbow trout may have come from stocking of fish from the south, this cannot be the explanation for its occurrence in the Erne.

Such inexplicable events are not that uncommon. The River Otter in south Devon has been the subject of parasitological investigations over a period of 15 years (Kennedy, 1996, 1997a). Until 1987, the only species of acanthocephalan present in eels *Anguilla anguilla* and brown trout *Salmo trutta* were *Acanthocephalus clavula* and *Neoechinorhynchus rutili*, both at low levels of infection (Kennedy, 1997a). In 1987, *P. laevis* was detected for the first time in eels (Table 8.6). and it subsequently increased in relative and absolute abundance until it came to dominate the helminth communities

Table 8.6. *The relative abundance of acanthocephalan species as a proportion (p_i)* *of the total number of helminths of all intestinal species in eels* Anguilla anguilla *in the River Otter*

Species	1985	1986	1987	1988	1989	1991	1994	1995	1996	1998
Acanthocephalus clavula	0	0	0.04	0	0	0.02	0	0	0	0
Neoechinorhynchus rutili	0	0.01	0	0	0	0	0	0	0	0
Pomphorhynchus laevis	0	0	0.05	0	0	0.07	0.19	0.19	0.16	0.29

Source: From Kennedy (1997a) (and updated), with permission.

of brown trout and become the commonest acanthocephalan in eels. How it arrived in the river and from where is unknown, although recently it has been identified as belonging to the Irish strain (O'Mahony *et al.*, 2004b) which is compatible with its use of trout as its definitive host in the river and which suggests it either arrived with a stray salmonid from Ireland or via an illegal introduction of fish from there. Whatever the source, it would not have been recognised as a successful introduction without the time series of samples. The same species of parasite has also appeared recently (in 1997), but independently, in the nearby River Culm (Table 8.7). This again appears to be a successful introduction as it now dominates the helminth communities of all the species in the river. Again the source of the introduction is unknown, although in this case it is known that chub *Leuciscus cephalus* had been illegally stocked into the river a few years previously and would seem to be the most likely source of the introduction. Both the rivers Culm and Otter contain high density populations of *Gammarus pulex*, the intermediate host of the acanthocephalans, and this together with the relative paucity of the other species of acanthocephalans in the rivers must have facilitated the introductions.

One of the best documented examples of a sucessful colonisation of a new locality by an acanthocephalan is that of *Paratenuisentis ambiguus* in the River Rhine (Taraschewski *et al.*, 1987). The acanthocephalan originated from North America, where it is a specific parasite of the American eel *Anguilla rostrata*. It uses *Gammarus tigrinus*, known from brackish waters off the east coast of the USA, as its intermediate host. The exact means of introduction of the parasite

Table 8.7. *The relative abundance of acanthocephalans expressed as a proportion (p_i) of the total numbers of all intestinal helminths of all species in eels* Anguilla anguilla *in the River Culm*

Species	Year			
	1981	1986	1997	1998
Acanthocephalus clavula	0	0	0.02	0.03
A. lucii	0	0.48[a]	0	0
Pomphorhynchus laevis	0	0	0.79[a]	0.87[a]
Echinorhynchus truttae	0	0	0.01	0

[a]Dominant species.

to Germany is not known. Native species of *Gammarus* were declining in the rivers Rhine, Weser and Elbe as a consquence of pollution by saline sewage and *G. tigrinus* was introduced to replace them in view of its tolerance of increased salinity. By 1995, *P. ambiguus* dominated not only the intestinal but also the total component communities of eels in the River Rhine (Sures *et al.*, 1999a). The other three species of acanthocephalans present, *Acanthocephalus lucii*, *A. anguillae* and *Pomphorhynchus laevis*, occurred at very low levels of prevalence and abundance. There were distinct site location preferences between species in the eel intestines, with *P. ambiguus* always occurring in the anterior third and *P. laevis* in the posterior third.

8.3 RARITY

As far as is known, no species of acanthocephalan is rare in the sense of being recognised as a rare species as defined by Gaston (1994). Rare species are considered here to be those that always exhibit low abundance and/or have small ranges. The eel specialist acantho- cephalan *Telosentis australiensis*, for example, is not considered a rare species even though it generally occurs at low prevalence and abund- ance (Kennedy, 1995) because it has a wide range along the east coast of Australia (possibly contiguous with that of its eel hosts) and a prevalence level of 40% has been recorded in one locality. It is necessary, however, to be very cautious here, as so many species have been recorded only once or at best on a very few occasions that it is impossible to determine whether they are actually rare or whether their apparent rarity reflects only a sampling artefact. When a species has just been introduced into a new locality it can be mistaken for being rare, but this will only be a temporary situation as the population is likely either to decline and the introduction fail, or increase in size until the threshold level for its survival is reached.

What is clear is that among the acanthocephalan species in fish in western Europe, which are one of the most intensively studied groups of acanthocephalans on the globe, no species is consistently rare or consistently common. The abundance ranges of each species may drop below unity in any locality but is likely to reach a maximum of below 100 in another. Where samples have been taken over a long time series as in the River Clyst (Kennedy, 1993b) or the River Otter (Kennedy, 1997a), it can be seen that species can be rare or absent over one period (*Neoechinorhynchus rutili* in the River Clyst) or common

Table 8.8. *Long-term changes in the relative abundance of each species of acanthocephalan as a proportion (p_i) of the total number of all acanthocephalans in eels* Anguilla anguilla *in two rivers*

Year	River Clyst				River Otter			
	N	A. clavula	N. rutili	P. laevis	N	A. clavula	N. rutili	P. laevis
1979	151	1.0	0	0	NS			
1980	149	1.0	0	0	NS			
1981	43	1.0	0	0	NS			
1982	61	1.0	0	0	NS			
1983	4	1.0	0	0	NS			
1984	0	0	0	0	NS			
1985	NS				0	0	0	0
1986	NS				1	0	1.0	0
1987	11	0.64	0.36	0	29	0.48	0	0.52
1991	8	0.87	0.13	0	5	0.2	0	0.8
1992	34	0.03	0.97	0	NS			
1994	8	1.0	0	0	3	0	0	1.0
1995	27	0.19	0.81	0	53	0	0.06	0.94
1996	NS				24	0	0	1.0
1998	32	0.66	0.25	0.09	50	0	0	1.0

NS, no sample taken.
Source: Data updated and from Kennedy (1993b) and Kennedy (1997a), with permission.

at one time but less common at others (*Acanthocephalus clavula* in the River Clyst) (Table 8.8). Similarly in the River Otter, *A. clavula* and *N. rutili* were always rare but *P. laevis* changed from rare to common (Table 8.8). In Lough Derg, *Acanthocephalus lucii* was always common and *A. anguillae* always rare (Kennedy & Moriarty, 1987, 2002). Here the two species shared the same intermediate and definitive hosts, and the most likely explanation of this situation was that the two species competed to the detriment of *A. anguillae* (pp. 153–4). In other localities *A. anguillae* can be dominant (Schabuss *et al.*, 2005). Whether these least common species persist in a locality at barely detectable levels of infection or whether they represent repeated unsuccessful invasions is not yet clear. However, the overall pattern that emerges is that rarity is essentially a local rather than a regional phenomenon: every species can be rare in one locality and abundant in another.

8.4 EXTINCTIONS

As far as is known, no species of acanthocephalan has become extinct globally, but Sprent (1992) drew attention to the dangers of this happening to parasite species. Although he focused on a group of nematodes, he stressed the dangers faced by all parasite species as a consequence of environmental changes, many of which were due to anthropogenic forces. These could result in habitat destruction and fragmentation, leading in turn to fragmentation of host populations and so to their decline. The smaller the population of hosts, and so parasites, the greater the stochastic probability of their extinction (MacArthur & Wilson, 1967). He felt that the parasites of any species in the Red Data Book were already in the vulnerable category and if their host(s) became extinct their parasites could disappear with them.

In their response to Sprent's seminal paper, Bush & Kennedy (1994) adopted a more optimistic approach. They focused on the dangers of parasite extinction due to anthropogenic fragmentation of host populations and in the absence of host extinction. At the infrapopulation level, the concept of extinction is trivial, as extinction is inevitable when the host dies. They considered that the extinctions of suprapopulations are unlikely in view of their complexity and the number of actual or potential host species that would have to be involved. They therefore focused on local extinctions of parasite metapopulations, particularly as a consequence of anthropogenic activities. A metapopulation is here defined as a system of local populations connected by dispersing individuals (Hanski & Gilpin, 1991). Local populations within a metapopulation may and do become extinct but they can be replaced by recolonisation from other local populations within the metapopulation, i.e. by a rescue effect. Metapopulations are often the key to a species persistence in a locality as they have a much longer time to extinction than local populations. As Price (1980) has emphasised, parasites normally live in patches, in which the probability of colonisation of a patch is low and the probability of extinction within a patch is high.

An example of such a local extinction and rescue effect is afforded by the population of *Acanthocephalus clavula* in eels *Anguilla anguilla* in a small river (Kennedy, 1984b). The River Clyst is a small tributary of the much larger River Exe, and the local population in the Clyst can be considered part of the metapopulation in the Exe catchment. The Clyst was subjected to management and canalisation in order to reduce flooding, and these changes resulted in a decline in the *A. clavula* population. Over a period of five years, prevalence levels fell from 60% to 2.0%, the mean intensity fell to below three and the frequency distribution changed from overdispersion to random until the population disappeared (Fig. 8.3). However, the population never actually became extinct: instead, it persisted at very low levels in

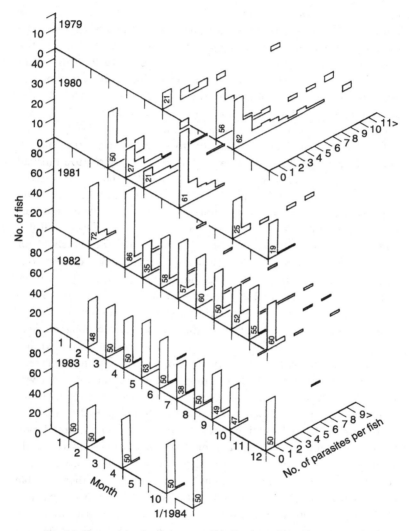

Fig. 8.3 Changes in the frequency distribution of *Acanthocephalus clavula* in eels *Anguilla anguilla* of the River Clyst. Small numbers within blocks are the number of fish examined. (From Kennedy, 1984b, with permission.)

subsequent years (Kennedy, 1993b) by virtue of a rescue effect from the River Exe components of the metapopulation.

Bush & Kennedy (1994) recognised that anthropogenic activities could result in host fragmentation. This could be advantageous in reducing the risk of extinction of a local population by providing a source for a rescue effect. However, it is more likely that resulting barriers between metapopulations could prove too difficult or too distant for hosts or parasites to overcome, and subsequently the host and/or parasite populations could decline to a level below the threshold at which they could persist (Anderson, 1982). This would be likely to pose more of a danger to autogenic species than to allogenic ones. The local extinction of the acanthocephalan species in Slapton Ley might be considered an analogous example of this. This small isolated lake is fed by four streams and empties directly to the sea. *Acanthocephalus lucii* was known to be present in the lake in both eels *Anguilla anguilla* and perch *Perca fluviatilis* up to 1975 but had disappeared by 1977, and *A. clavula*, also found in both host species, disappeared before the winter of 1984–5 (Table 8.2, 8.3). The reason for the disappearance of *A. lucii* is not known, but *A. clavula* populations had been declining for some time. Over the winter of 1984–5 all fish populations crashed in early 1985 to a level below the threshold for transmission of most parasites (Kennedy, 1998). The metapopulations of both autogenic species had become locally extinct, and the isolation of the lake prevented any recolonisation from the nearest metapopulation some 20 km distant. Although anthropogenic habitat fragmentation was not involved in this example, anthropogenic activities were, as the crash of the fish populations resulted from winterkill, which in turn was due to hyper-eutrophication within the lake.

A feature of the potential danger of local extinctions that was stressed by Bush & Kennedy (1994) was that parasite populations exhibited several hedges against extinction. When applied to acanthocephalans, high reproduction rates could often enable a population to recover from reduced population levels and rescue effects could achieve recolonisation. Wide specificity, a characteristic of generalist acanthocephalans, could also enable them to utilise other suitable host species should any preferred host species become locally extinct. (Fig. 8.4). The ability of aquatic acanthocephalans to use migratory species of fish to effect colonisations (Lyndon & Kennedy, 2001) can also be viewed as a hedge against extinction. As long as the parasite can survive in the colonisation host, it can await the return to, or appearance of, a more suitable host in the original or new locality.

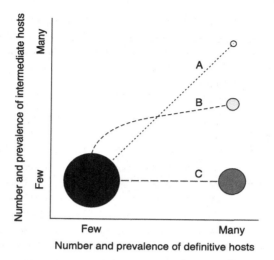

Fig. 8.4 Host specificity as a hedge against extinction: hypothetical probability for the extinction of a parasite metapopulation. The larger and darker the sphere, the greater the probability of becoming extinct. (Reprinted from Bush & Kennedy, 1994, copyright 2004, with permission from Elsevier.)

Even accidental hosts may be significant in avoiding extinction. The eel *Anguilla anguilla* is normally regarded as an unsuitable and accidental host of *Pomphorhynchus laevis* as the parasite is believed to be unable to become sexually mature in eels and is only acquired by them incidentally when they feed on infected gammarids. However, *P. laevis* is able occasionally to reproduce in eels (Table 8.9) and so the eel could serve as a colonisation host and an escape from local extinction for this species. This may also be true for other acantho-cephalan species in what are generally regarded as accidental or unsuitable hosts.

Some hedges against extinction such as progenesis are not available to acanthocephalans, but their ability to use common, widespread and ecologically tolerant species of arthropods as intermediate hosts must contribute to their ability to avoid local extinctions. Although acanthocephalans are generally *r*-selected, little is known about their evolutionary rates. Laboratory infections of laboratory strains and wild type *Moniliformis moniliformis* in rats show little evidence of any change over 60 generations (Reyda & Nickol, 2001). Host captures have also been suggested by Bush & Kennedy (1994) as a hedge against extinction, and such captures over evolutionary time have been

Table 8.9. *Maturation of* Pomphorhynchus laevis *in eels in Britain*

Locality	Year	No. of eels examined	Prevalence (%)	Abundance (mean)	Maximum intensity	% females gravid
R. Avon	1987	24	66.7	8.8	52	0
	1993	18	94.4	8.6	37	0
R. Culm	1997	32	62.5	10.9	88	0
	1998	13	69.2	11.4	39	0
	2000	25	56.0	3.8	36	0
R. Otter	1995	22	45.4	1.3	18	0
	1997	30	26.7	1.7	21	0
	1998	22	22.7	2.1	29	4.5
R. Thames	1998	16	16.0	1.0	9	54.5

Source: Data from Kennedy (1996, 2001), Lyndon & Kennedy (2001) and Kennedy (unpublished).

suggested for the acquisition of *Corynosoma* and *Bolbosoma* species by marine mammals from birds (p. 31) and for the capture of *Plagiorhynchus cylindraceus* from birds by marsupials (p. 169). Over a shorter timescale, the use of brown trout *Salmo trutta* and *Gammarus duebeni* as preferred definitive and intermediate hosts respectively by *Pomphorhynchus laevis* in Ireland has been considered as host captures following the introduction of the parasite to Ireland, from which its normal species of both definitive and intermediate host are absent (Kennedy *et al.*, 1989; Bush & Kennedy, 1994).

It appears, therefore, that almost all the hedges against local extinction identified by Bush & Kennedy (1994) are available to acanthocephalans. The same hedges must also minimise the probability of extinctions of suprapopulations. Global extinctions are a different matter as no group of parasites has any protection against cataclysmic events. However, the wide specificty of adult acanthocephalans and their use of common and widespread invertebrate species as inter-mediate hosts must make their survival more probable even if a preferred species of host disappears from the Red Data Book by extinction. This is not to suggest that extinction of an acanthocephalan species cannot happen: it clearly can and indeed over time all species become extinct. What is evident is that local extinctions, whether due to natural events or anthropogenic activities, are not normally a matter for particular concern.

9

Relations to ecosystem changes

9.1 BIOLOGICAL TAGS

Parasites are integral components of the ecosystem in which they live. They are not merely superimposed on food webs but they flow through them (Marcogliese & Cone, 1997a,b; Marcogliese, 2002). Any effects that they have on their host populations, invertebrate and vertebrate, can affect the ecosystem. Because of this intimate relationship, they can also be affected by any changes in the free-living components of the ecosystem. Under some circumstances, therefore, they can provide information on the condition and functioning of an ecosystem (Marcogliese & Cone, 1997b). In particular they may be able to provide information on the biology of some of the species within it, especially on their movements and diets, and on changes that have taken place in that ecosystem. They can in effect serve as biological tags and as indicators of the state of an ecosystem.

The use of parasites as biological tags to indicate the geographical origin of a host is particularly applicable to marine fish. Margolis (1963, 1965) was among the first people to fully appreciate the value of parasites in this context; he demonstrated that it was possible to distinguish the country of origin of Pacific salmon of the genus *Oncorhynchus* caught at sea by their parasite fauna: some species of parasites were found only in fish from North America and others only in fish from Russia. MacKenzie (1983, 1986, 2002; Williams *et al.*, 1992) has also used parasites exensively as tags to distinguish different stocks of a fish species and to detect the range and extent of migrations of stocks. MacKenzie (1986) has laid out the guidelines that need to be followed in selecting biological tags. These include: (a) knowledge of the complete life cycle of the parasite, (b) site reproduction within the host, (c) simplicity in location and identification of the parasite,

(d) an absence of any pathological or behavioural effect of the parasite on the host being studied, (e) longevity, (f) relative abundance of the parasite, and (g) geographic variation in levels of intensity of infection. Few acanthocephalans meet these conditions: as many life cycles are unknown, many species are very difficult to identify, most affect the behaviour of their intermediate hosts and all are short lived. Williams & Jones (1994) have summarised the studies using parasites in separation of fish stocks, in studying changing patterns in stock migrations and in studying recruitment and seasonal migrations of fish, and it is clear that while acanthocephalans may be members of characteristic communities that can be used as tags, by themselves they are seldom of value.

Comiskey & MacKenzie (2000), however, did suggest that *Corynosoma* species might be useful biological tags for saithe *Pollachius virens* in the northern North Sea. Juvenile acanthocephalans were found in 19% of juvenile saithe from Scottish grounds, whereas fish from Norwegian nursery grounds were unaffected by these parasites. Pippy (1969) considered the possibility of using *Pomphorhynchus laevis* as a biological tag for Atlantic salmon *Salmo salar*. He found the species to be common (53% overall) in smolt runs from Ireland and very uncommon in smolt runs from England and Scotland (Fig. 9.1), but he failed to find the American species *P. bulbocolli* in smolts from North America. Neither species was found in salmon captured off West Greenland (Pippy, 1980). He therefore concluded that the parasite did not really meet the key conditions for use as a tag: it could not be used to indicate the continent of origin of adult fish or the river of origin of smolts, although it might have some limited value in distinguishing smolts of Irish and English origin in coastal waters. The only acanthocephalan found in adult salmon was *Echinorhynchus gadi*, but that is a marine species and thus was of no value as a tag. Gibson (1972) came to a similar conclusion in his attempt to use *P. laevis* as a tag for the river of origin of flounders *Platichthys flesus* in north-east Scotland.

Balbuena & Raga (1994) attempted to use *Bolbosoma capitatum* as one of four species as indicators of segregation and social structure of pods of long-finned pilot whales *Globicephala melas* off the Faroe Islands. They sampled seven pods, and found that the abundance of *B. capitatum* in the different pods supported other evidence from previous studies in suggesting that there was a degree of segregation between the pods but the evidence was not strong enough to confirm the existence of separate stocks.

Fig. 9.1 Distribution and abundance of *Pomphorhynchus laevis* in *Salmo salar* smolts in the British Isles. Black areas indicate percentage of smolts infected in each locality. (From Pippy, 1969, with permission.)

However, acanthocephalans can be used as a tag on a more local scale. Cribb *et al.* (2000) studied populations of the fish *Lutjanus carponotatus* on the Great Barrier Reef. Samples were taken from two sites only 300 m apart: one from the reef flat in shallow waters and one from the reef slope in deeper waters. The mean (and range) abundance of *Pomphorhynchus heronensis* was 2.6 (0–9) at the slope locality but 39.6 (1–122) at the flat (Table 9.1). Neither year, season nor size of host contributed significantly to this difference and so they concluded from the differences in the abundance of the acantho-cephalans that transmission of the parasite was very local and that the fish had limited local movement. Kennedy & Burrough (1978) were able to use *Neoechinorhynchus rutili* as an indicator of the origin of brown trout *Salmo trutta* in a small lake and as a means of distinguishing native and introduced trout. Infection of trout appeared to take place

Table 9.1. *Pomphorhynchus heronensis in Lutjanus carponotatus at Heron Island on the Great Barrier Reef*

Location	No. of fish	Parasite			
		Prevalence (%)	No. with 0	Mean intensity	Range of intensity
Reef slope					
North Wistari	11	54.5	5	1.8	0−5
Heron Channel	23	69.6	7	2.9	0−9
North Heron	2	100.0	0	4.0	3−5
Pooled reef slope	36	66.7	12	2.6	0−9
Reef flat	30	100.0	0	39.6	1−122

Source: Data from Cribb *et al.* (2000).

only in the streams feeding into the lake and levels could attain 66% in older trout which had entered the stream to spawn: the parasite was never found in stocked fish of any age.

A slightly unusual use of an acanthocephalan as a tag occurred in a court case of suspected poaching (Kennedy, unpublished). A sample of eels *Anguilla anguilla* was recovered from nets in a small lake and the fishermen concerned claimed that the eels had been caught in a nearby estuary and had been introduced into the lake for an hour or so only to refresh them before their journey to market. However, the fish were infected with mature freshwater *Acanthocephalus clavula*, indicating that they had been in fresh water for several weeks, not just an hour, and the only site of infections of *A. clavula* in that region was the lake in which the eels were found. The fishermen were convicted on this evidence.

Although it does not come strictly under the term tag, acanthocephalans can be used to provide information on host diets. Thus the presence of *Acanthocephalus* species in a host indicates that the host has fed on benthic invertebrates including species of *Asellus* (Kennedy *et al.*, 1992). The presence of *P. laevis* in a fish in Europe similarly indicates that the host has fed on a species of *Gammarus*, and the presence of *P. bulbocolli* in North America and *P. patagonicus* in Argentina that the hosts have fed on a species of *Hyalella*. The abundance of the acanthocephalan in the host individuals can also provide some indication of the extent to which the host has fed on the invertebrate: for example, the high levels of infection of eels

Anguilla anguilla with *Acanthocephalus lucii* in Meelick Bay in Lough Derg in Ireland suggest that eels there normally fed intensively on *Asellus aquaticus* (Kennedy & Moriarty, 2002), and a decline in the acantho- cephalan levels in the bay (Table 6.3) suggested that changes in the movement patterns of eels in and out of the bay were taking place. Similarly, the extent to which different species of fish are infected with *Pomphorhynchus laevis* in the River Avon (Table 4.4) or *Metechino- rhynchus salmonis* in Cold Lake (Table 4.5) is an index of the extent to which they have fed on *Gammarus pulex* or *Pontoporeia affinis*.

9.2 ACANTHOCEPHALANS AS INDICATORS OF POLLUTION

9.2.1 Organic pollution

The issue of whether parasites can be used as indicators of organic pollution is a controversial one. There is general agreement that organic pollution can affect parasite infection levels, but there is less agreement over whether parasite infection levels can indicate organic pollution. There are several conditions that have to be met before parasites can be used as pollution indicators or monitors, and these have been set out clearly by Overstreet (1997) and by Williams & MacKenzie (2003). Ideally, a host and its parasites need to have a restricted home range, the host needs to be infected with several parasite species that utilise a variety of hosts in their life cycles and the host needs to be easily sampled. As Overstreet has also pointed out, it is desirable to use the total parasite community and it is necessary to have an expected parasite community against which to measure any changes. This is possible if samples are taken over a period of time and in the same season, but it is difficult to find reference locations to compare samples over space at any one time because of the high local degree of variation in parasite community composition and structure. Because diversity of free-living communities tends to decline with organic pollution, it might be predicted that parasite communities will also be less diverse as potential hosts decline or disappear, but this may be counteracted by the increased abundance of some species of hosts that are more pollution tolerant. Thus, species of *Gammarus* may be replaced by species of *Asellus* as levels of organic pollution increase and so one might predict that species of *Pomphorhynchus* and *Echinorhynchus* might be replaced by species of *Acanthocephalus*.

All these conditions and reservations apply to helminth parasite communities which include acanthocephalans, and it is difficult to consider the acanthocephalans separately from other members of the community in view of the desirability of using the whole helminth community to indicate pollution. However, occasionally the impact of organic pollution on acanthocephalan species has been singled out. Kussat (1969) reported that prevalence and intensity of *Octospinifer macilentus* and *Neoechinorhynchus cristatus* were higher in catfish *Catastomus commersoni* upstream of the inflow of effluents of domestic and industrial wastes into the Bow river from Calgary than in fish downstream. This was attributed to the decline and disappearance of amphipods and ostracods from the polluted regions of the river. Overstreet & Howse (1977) similarly reported that the prevalence and intensity of infections of *Dolfusentis chandleri* in the Atlantic croaker *Micropogonias undulatus* in the Gulf of Mexico declined as pollution levels increased. They believed that this was due to a decline in the numbers and availability of the amphipod intermediate host. Vidal-Martinez *et al.* (2003) found *Gorgorhynchus bullocki* in *Ariopsis assimilis* in four localities in Chetumal Bay, Mexico, and although infection levels were low, they were higher at the least polluted site. MacKenzie *et al.* (1995) also considered that acanthocephalans in themselves contributed little to an understanding of effects of pollution on marine helminths, as they generally formed only a minor part of the community richness or structure of helminths in fish. Isolated studies, such as their finding that exposure to water-soluble fractions of oil reduced prevalence and intensity of *Echinorhynchus gadi* in cod *Gadus morhua*, were at best only indicative.

Gelnar *et al.* (1996) investigated the relationship between helminth community diversity and pollution in the Morava river, a tributary of the River Danube. They examined communities in chub, *Leuciscus cephalus*, at two localities, which were considered to be comparable except that one was clean with a saprobic index of 2.16 and one was polluted with an index of 3.15. The helminth communites were rich, but rather less so in the polluted site. Only two acanthocephalans were present and in both localities: *Pomphorhynchus laevis* was 11th in rank abundance and *Acanthocephalus anguillae* 15th at the polluted site, with corresponding values of 32nd and 26th for the clean site. Both prevalence and abundance levels of the two species were low, and considered in isolation from the other helminth species they provided no indication of pollution. Galli *et al.* (1998) undertook a somewhat similar study of parasites of chub in four otherwise comparable river reaches characterised by different levels of pollution. They focused

specifically on acanthocephalans, reporting on the same two species
P. laevis and *A. anguillae*. They found that *P. laevis* was restricted to
the unpolluted and slightly polluted sites (Fig. 9.2), whereas prevalence
of *A. anguillae* increased significantly, though intensity only slightly,
with increasing pollution. They believed that these differences were
related, at least in part, to the changes in their intermediate hosts,
Echinogammarus stammeri and *Asellus aquaticus* respectively, with pollution: the latter species being more tolerant of polluted waters. There
is also some evidence that gammarids infected with acanthocephalan
cystacanths are more vulnerable to pollutants, as Brown & Pascoe
(1989) found that *Gammarus pulex* infected with *P. laevis* were more
vulnerable to pollution and especially if inorganic metals such as
cadmium were present.

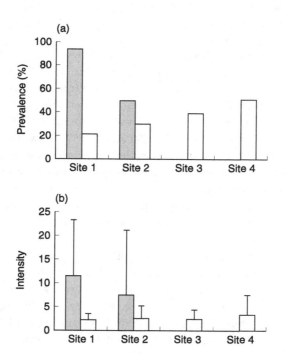

Fig. 9.2 Prevalence (a) and intensity (b) values for *Pomphorhynchus laevis*
(shaded bars) and *Acanthocephalus anguillae* (white bars) in four rivers
experiencing different levels of organic pollution. Site 4 is the most
polluted and site 1 the least. Vertical lines indicate standard error.
(From Galli *et al.*, 1998, with permission.)

The effects of an effluent from a paper mill, which resembles that of organic pollution in many respects, on helminth communities in the winter flounder *Pleuronectes americanus* were examined by Khan & Payne (1997). They compared samples taken near the mill and at reference sites. Flounders near the mill had higher infection levels of *Echinorhynchus gadi* than those taken at the more distant sites. However, they believed that levels of the acanthocephalan alone could not provide a clear indication of the effects of the pollution: it was necessary to consider them as part of a community and to study the whole community. Valtonen *et al.* (1997) have also studied the impact of pulp mill effluents on helminth communities in fish from central Finland. They determined that the effluents had an effect on some members of the communities, but little detectable effects upon the acanthocephalans. When they later (Valtonen *et al.*, 2003) studied the communities as indicators of recovery from pollution, they again found that only a few species, and not the acanthocephalans, provided evidence for helminth community recovery.

The general conclusion that can be drawn from all these studies is that parasite communities do have the potential to provide pollution indices in the same way that chemical and physical data do. Parasites with complex life cycles and intermediate hosts which are vulnerable to pollution are in some respects ideally suited to be useful indicators of pollution (McVicar, 1997). However, the real problem is that natural spatial and temporal variation in helminth community composition is so extensive that it may be very difficult indeed to detect changes that are due to pollution. It will be necessary somehow to reduce or understand this natural variation before the potential of parasites, including acanthocephalans, as indicators can be realised.

Most of the effects of organic pollution appear to be mediated through changes in the abundance of free-living arthropod intermediate hosts, and it is far simpler to use these directly as indicators of pollution, as is currently often done. Kennedy (1997b) urged great caution in the interpretation of data sets that claimed to link helminth changes to pollution levels and/or tried to use helminth parasites as indicators of organic pollution. He questioned whether there was any real advantage in trying to use parasites when there were already many excellent indices of organic pollution that made use of the far better studied and understood free-living species, and when it was yet to be demonstrated that parasite-derived indices would be in any way superior to these.

9.2.2 Acidification

The most detailed studies of the effects of acidification and lowered pH on helminth communities have been undertaken in Nova Scotia on parasites of the American eel *Anguilla rostrata*. The earliest study (Cone *et al.*, 1993) showed clearly that parasite communities were less rich in the acidified waters but there were no acanthocephalans present in the localities sampled. A later study (Marcogliese & Cone, 1996) confirmed that helminth diversity decreased as pH also declined and this study did include sites in which acanthocephalans were present. *Pomphorhynchus bulbocolli* was found in only one river, and at a prevalence of 28.5% and mean intensity of 28.5, and the river had a pH > 5.4. *Echinorhynchus lateralis* was found in only four rivers but attained highest levels of abundance in a river where pH ranged between 5.1 and 5.4 and *E. salmonis* was also found in only four rivers, but where pH values were > 4.7 and ranged to > 5.4. The distribution of *P. bulbocolli* was restricted to that of its preferred definitive host species, which is not the eel, and that of *E. salmonis* to that of a more suitable definitive host, the brook char *Salvelinus fontinalis*, which is very acid-sensitive. The distribution of *E. lateralis* was not readily explicable, but may relate to that of its intermediate host. The three species showed local and disjunct distributions, occurred at low levels of infection and contributed very little to community diversity.

9.2.3 Heavy metal pollution

By contrast with the above findings, the relationship between heavy metal pollution and acanthocephalans is very clear. We now know, thanks very largely to the work of Sures and his colleagues (reviewed by Sures *et al.*, 1997a, 1999b and Sures, 2003, 2004) that acanthocephalans are excellent indicators of the presence and levels of heavy metals in an ecosystem. Sures *et al.* (1994) analysed chub *Leuciscus cephalus* and its parasite *Pomphorhynchus laevis* taken from the River Ruhr for lead and cadmium. They reported that the mean lead level in the parasite was 284 times more than that in the host intestine, 771 times more than that in the host liver, 27 000 times more than that in host muscle and 11 000 times higher than in the water of the River Ruhr (Fig. 9.3). They thus concluded that *P. laevis* may serve as a very sensitive bioindicator for the presence of biologically available lead in aquatic ecosystems.

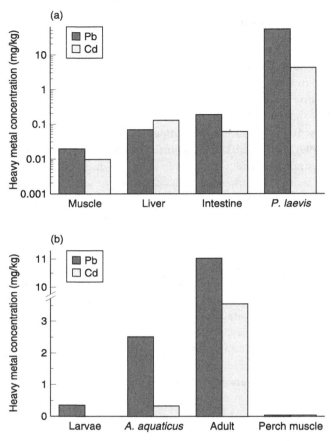

Fig. 9.3 Lead and cadmium concentrations (wet weight) in acanthocephalans from the River Ruhr. (a) In different organs of chub *Leuciscus
cephalus* and in *Pomphorhynchus laevis* and (b) in muscle of perch *Perca
fluviatilis*, in the amphipod *Asellus aquaticus* and in cystacanths and adults
of *Acanthocephalus lucii* in perch. (From Sures *et al.*, 1997a, with permission
of the authors.)

Further studies have reinforced and extended this conclusion.
Sures (2002), Sures & Siddal (1999, 2003), Sures *et al.* (1994) reported that
this was not just a property of *P. laevis*, as both *Acanthocephalus lucii* in
perch *Perca fluviatilis* and *Paratenuisentis ambiguus* in eels also contained
far higher levels of lead than did the tissues of perch or the river water.
Sures & Taraschewski (1995) confirmed that levels of cadmium were
also bioaccumulated more effectively by parasites than by their fish
hosts. They found that *P. laevis* harboured 31 times more cadmium

than did the liver of chub, 67 times more than the intestine and 400 times more than the muscle of chub. Similarly, the levels of cadmium accumulated by A. lucii were 3 times higher than in the liver of perch, 18 times higher than in the intestine and 133 times higher than in perch muscle. By contrast, larvae of A. lucii taken from Asellus aquaticus accumulated less cadmium and lead than did the isopod itself, mean values for which were 16 times higher for lead and 7 times higher for cadmium compared with the values recorded from the cystacanths (Fig. 9.4). Siddal & Sures (1998) confirmed that cystacanths of P. laevis in Gammarus pulex also did not accumulate lead, and experimental infections revealed that significant bioconcentration by P. laevis was already apparent within a week post infection. It later transpired (Sures & Siddal, 1999; Sures et al., 2003) that infected chub contained less lead in their intestinal tissues than did uninfected ones, indicating that P. laevis was acting as a lead sink for chub. The parasite was absorbing lead from the bile and thus reducing the absorbtion of lead by the intestinal wall. The implication of this was that measurements of levels from host tissues may show lower levels of heavy metals if acanthocephalans are present. It also became apparent (Sures et al., 1999b) that other species of acanthocephalan, including the introduced Paratenuisentis ambiguus in eels, could also bioaccumulate metals but nematodes could not.

Acanthocephalans have indeed proved to be very sensitive indicators of heavy metal pollution. They are able to bioaccumulate a wide range of metals (Fig. 9.4), some more effectively than others (Sures et al., 1997a). Even more significantly in relation to bioindication of heavy metals, they accumulate lead and cadmium more effectively than does the free-living freshwater mussel Dreissena polymorpha (Fig. 9.4), which was hitherto considered to be the most sensitive bioindicator of heavy metals. Indeed, in these samples taken from Lake Mondsee the highway site was close to a motorway and received road run-off via a small stream entering the lake (Sures et al., 1997b). The parasite was thus actually recording heavy metal pollution from a terrestrial source. It has subsequently been demonstrated (Sures et al., 2003; Zimmermann et al., 2005) that P. ambiguus can take up noble metals originating from ground catalytic exhaust gas converters of cars which accumulate in sediments of aquatic ecosystems (Fig. 9.5) and so serve as a very sensitive indicator to assess the degree of environmental contamination from this source. Uptake of metals is possible over a wide range of water chemical conditions, and whereas salinity reduces the lead burden in host tissues, acanthocephalans

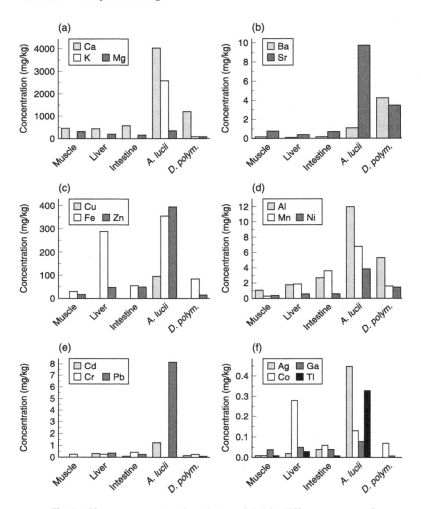

Fig. 9.4 Element concentrations (wet weight) in different organs of perch *Perca fluviatilis*, in *Acanthocephalus lucii* in perch and in the bivalve mollusc *Dreissena polymorpha* from Lake Mondsee. (From Sures *et al.*, 1997a, with permission of the authors.)

are unaffected by it (Sures *et al.*, 2003). Acanthocephalans may also have a role to play in the assessment of pollution in remote areas such as the Antarctic, as levels of many heavy metals in the acanthocephalan *Aspersentis megarhynchus* in fish from the South Shetland Islands have been shown to be accumulated and higher than those in their host tissues (Sures & Reimann, 2003).

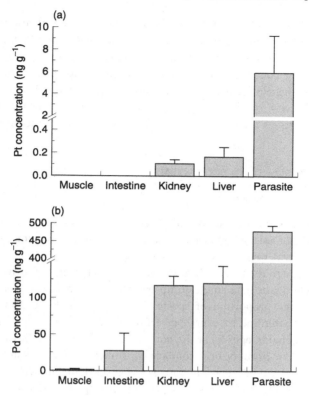

Fig. 9.5 Distribution of platinum (a) and palladium (b) in the tissues of the eel *Anguilla anguilla* and in *Paratenuisentis ambiguus* after six weeks of exposure to automobile catalytic converter material. Values are mean concentrations and error bars show standard deviation. (From Zimmermann *et al.*, 2005, with permission.)

Accumulation has now also been demonstrated in *Macracantho-rhynchus hirudinaceus* in pigs, in which some metals are bioconcentrated up to 160-fold (Sures *et al.*, 2000a) and in experimental infections of rats with *Moniliformis moniliformis*, in which cadmium and lead can be bioconcentrated up to 119-fold (Scheef *et al.*, 2000; Sures *et al.*, 2000b). These studies show that acanthocephalans can also serve as indicators of metal pollution in terrestrial systems, but their utility may be limited by the fact that the acanthocephalans are not as common or abundant in terrestrial hosts as in aquatic ones.

Acanthocephalans such as *P. laevis* have been used to detect heavy metal pollution in rivers such as the Danube (Thielen *et al.*, 2004), where *P. laevis* in barbel *Barbus barbus* were analysed for 21 metals.

Ten of these were found in significantly higher concentrations than in host tissues, and eight of them were accumulated even better than in established bioindicators such as *Dresissena polymorpha*. The results confirm the conclusions of an earlier study by Schludermann *et al.* (2003) on the same parasite and host in the same river. They confirm the sensitivity of acanthocephalans as indicators of heavy metals and as sentinels for metal pollution in natural aquatic systems.

9.3 IMPACT ON ECOSYSTEMS

9.3.1 Impact of native species

Parasites are an integral part of all ecosystems. Superimposed on the food webs of free-living species is a web of parasites and for many of them survival relies on their moving up the food web in a series of stages through the trophic levels until they reach their definitive hosts, and down again in a single jump by release of eggs. Since parasites form a large proportion of species on the planet (Price, 1980) they have the potential to exert powerful effects on the other species. Acanthocephalan eggs lie at the bottom of the food chain; larval stages are passed in primary or secondary consumers and adults are found in predators at or near the top. An acanthocephalan will typically move up the web in two stages via the arthropod intermediate host and vertebrate definitive host, but it may take as many as four stages if a vertebrate paratenic host is incorporated into the life cycle and if post-cyclic transmission takes place. At each stage, part of the acanthocephalan population may spread laterally and become located in a wide variety of host species. Thus, each species may ramify through the food web both vertically and laterally, and if several acantho-cephalan species are present in the ecosystem the ramifications may be very complex and involve a large number of free-living species. Acanthocephalans, because they can affect the encounter rates between predators and prey, have the potential to influence the structure of, and energy flow through, the whole food web and may even assist in stabilising ecosystems (Combes, 1996, 2001).

Every acanthocephalan species has the potential to make a powerful impact on its ecosystem by virtue of its impacts on its host species at every level. As Holmes (1982) and Holmes & Price (1986) have pointed out, parasites may: (a) have lethal or sublethal effects on their hosts, the effects of which on the component population of hosts will depend on the degree of aggregation displayed by the parasite, (b) regulate the host population through parasite-induced host mortality if such mortality is additive, (c) act as a selective agent determining which host individuals will be killed if mortality is compensatory, (d) influence the co-existence of species by acting as a 'weapon of competition' and (e) influence the diet of host species and so prey–predator interactions in the ecosystem. This can be achieved by (i) reducing the energy cost of prey capture by modifying intermediate host behaviour, (ii) rerouting the transfer of energy and material between species, (iii) concentrating predation on individual hosts, which may themselves be of lower reproductive value as a consequence of the parasite, and (iv) influence patchiness of dispersion.

The effects of a parasite on its host and vice versa can be viewed as a continuous variable between extreme susceptibility on the part of the host and pathogenicity on the part of the parasite, to total resistance (Toft, 1991). If parasites do regulate their host populations, this would have wide consequences for community structure, food webs and energy flow. Ecologists, however, have tended to ignore the possible impact of parasites on ecosystems. Parasite-induced host mortality may have a negative effect on the parasite population if the parasites are taken out of the system, but it may have a selective effect on the host. Parasites, therefore, can affect α-diversity, the length of a food chain and γ-diversity. The role of parasites in influencing co-existence or competitive exclusion of free-living species in a community may also be important and parasite-mediated competition significant. 'A non-specific parasite is a powerful competitive weapon' (Hudson & Greenman, 1998), as it may influence the outcome of competition if one species achieves success by transmitting a pathogen to a more vulnerable competitor.

Acanthocephalans are only a small phylum with a small number of species, but their ability to infect a wide range of host species in a wide range of ecosystems means that their potential influence is out of proportion to the number of species. However, many of these potential effects are unlikely to be realised at the definitive host level because they depend on the pathogenic effects of the parasite on its hosts, and acanthocephalans rarely have deleterious effects on their definitive hosts. Acanthocephalans may occasionally cause mortality of heavily infected hosts (Thompson, 1985a,c), but this is the exception and it may have little or no long-term impact on the host population. Acanthocephalans may compete with other acanthocephalan species or species of other parasitic phyla within definitive hosts, but this appears to have no impact on the host populations and as yet there are no examples of acanthocephalans being used as a competitive weapon.

The real impact of acanthocephalans on ecosystems is evidenced at the intermediate host level (Marcogliese, 2002; Mouritsen & Poulin, 2002). As stated earlier (Chapter 5), almost every species of acanthocephalan is able to modify the behaviour of its intermediate host in an adaptive manner that facilitates transmission of the infected individual to the definitive host. Moreover, the species utilised as an intermediate host is often a keystone species in its community: for example, species of *Gammarus* and *Asellus* are often key species in freshwater communities, species of crabs in intertidal communities and species of roaches in terrestrial ones. The effect of acanthocephalans may be as a direct cause of host mortality or by impacting on host behaviour. Parasite-induced host mortality may have a serious impact on an intermediate host population, as has been demonstrated by Latham & Poulin (2001, 2002) in the case of mud crabs infected by an acanthocephalan (Fig. 5.3). Preferential manipulation of one intermediate host species over another could influence diversity within the ecosystem. Moreover, any decline in one crab population may cause predators to switch their attention to alternative prey species with a series of possible consequences for the whole food web that are discussed by Mouritsen & Poulin (2002). If, however, parasite-induced intermediate host mortality favours transmission, a number of predator species may switch to feeding on the infected intermediate hosts because it is energetically effective to do so. They are easier to catch, and so the predator expends less energy in catching them. For example, in the River Avon, every species of fish feeds more or less extensively on *Gammarus pulex* because it is infected with cystacanths of *Pomphorhynchus laevis*; levels of infection may be higher in species such as the grayling *Thymallus thymallus* (Table 4.4), a totally unsuitable definitive host, than in the suitable hosts barbel *Barbus barbus* and chub *Leuciscus cephalus* (Hine & Kennedy, 1974a). Similarly, all the species of fish in Cold Lake and Lake Michigan predate heavily on infected *Pontoporeia affinis* (Table 4.6 and p. 64) because it is easier to catch. In Ireland, situations have been reported where an intermediate host is so common in a locality that fish feed preferentially on it and so acquire acanthocephalan species that do not normally infect them: for example, the exceptionally high levels of *Acanthocephalus clavula* in trout *Salmo trutta* in an Irish lake is not an example of host switching but is an example of the trout feeding heavily and selectively on *Asellus meridianus* (Byrne *et al.*, 2004), and the abundance of *Acanthocephalus lucii* in eels in Lough Derg is a direct consequence of their feeding selectively on infected *Asellus aquaticus*. Almost certainly such effects are widespread, but it is often difficult to assess whether this potential effect on food webs is realised because few studies have been designed with this aim and the range of systems is so wide.

9.3.2   Impact of introduced species

Stable ecosytems and food webs may be so complex that it is almost impossible to detect the effects of a parasite even if it does impact on a key species (Marcogliese, 2002). However, it should be easier to observe a parasite's impact when it is introduced naturally or by anthropogenic activities into an ecosystem, and especially if the system is perturbed by the parasite. Introduction of parasites into a new system has always been a matter of concern, to the extent that Torchin *et al.* (2002) consider introductions to be second only to habitat loss as a threat to biodiversity. The impact of an invader relates to its biomass and success and as introductions often attain high densities due to better resources, reduced competition and paucity of enemies (parasite or predator) the threat is very real. Invading hosts, however, often lack parasites from their place of origin and the richness of their parasites is often less than half of that in their natural habitat (Torchin *et al.*, 2003). The impact of an invader may, however, be much wider than paucity of parasites may suggest.

No species of acanthocephalan has been shown to have success-fully invaded the freshwater fish fauna of the British Isles since the post-glacial era (Kennedy, 1993c, 1994), although in recent times species of monogenans, digeneans, cestodes and nematodes have done so. This may mean no more than that the fish acanthocephalan fauna is already similar to that of the rest of western Europe, i.e. all the local species have already invaded, or it may mean that any invasion has not been successful or detected as such. Whatever the reason, it is necessary to look at individual localities to see any impact made by an invading acanthocephalan. Examples of such invasions have been given earlier (Tables 8.6 and 8.9) of *Pomphorhynchus laevis* successfully invading the rivers Otter and Culm. In neither case was there any detectable impact of the parasite on the fish or existing species of acanthocephalans in the rivers. In the river Otter, levels of *Acanthocephalus clavula* and *Neoechinorhynchus rutili* were already very low in fish, and *P. laevis* appears to have moved into a vacant niche without any impact on the existing acanthocephalan community or on the fish populations. The situation in the River Culm was remarkably similar in that the native species of acanthocephalans persisted at low levels and the invading *P. laevis* moved into a vacant niche. As stressed previously in the context of natural ecosystems, because acanthocephalans have no pathogenic effect on their definitive hosts they have little or no impact when they are in the fish. The impact again results from their effects on the intermediate host: in the River Otter the parasite was found at high prevalence levels in almost every species of fish (Table 9.2) and the same was true in the River Culm (Kennedy, unpublished). Here again, the acanthocephalan had a major impact on the diets of the fish and so on the food web through its effects on the intermediate hosts. There appears also to be little or no evidence from marine ecosystems that invading acanthocephalans have any impact through their definitive hosts (Torchin *et al.*, 2002).

There are, however, two particularly well documented examples of the close relationships between acanthocephalans and introduced

Table 9.2. *The colonisation of the River Otter by* Pomphorhynchus laevis *in three hosts*

		Anguilla anguilla				Platichthys flesus		
Year	No.	Prevalence (%)	Abundance (± SD)	Range	No.	Prevalence (%)	Abundance (± SD)	Range
1985	43	0	0	0	6	0	0	0
1986	35	0	0	0	0	0	0	0
1987	188	0	0	0	0	0	0	0
1988	85	8.2	0.18	0−5	65	3.1	0.03 (0.2)	0−1
1989	10	0	0	0	33	9.1	0.42 (1.7)	0−9
1991	16	12.5	0.25 (0.8)	0−3	12	0	0	0
1994	10	10.0	0.3 (0.9)	0−3	55	43.6	2.4 (6.7)	0−37
1995	75	22.6	0.6 (2.3)	0−18	55	43.6	1.2 (2.3)	0−11
1996	17	23.5	0.23 (0.8)	0−1	NS			
1997	17	17.6	0.18 (0.4)	0−2	NS			
1998	22	27.2	2.13 (9.4)	0−29	NS			
Trout								
1995	12	100.0	23.8 (16.9)	2−54				
2001	5	100.0	28.0 (18.2)	1−75				

NS, no sample.
Source: From Kennedy (1996, 1997a) and unpublished.

species. The first of these comes from the River Rhine. Taraschewski *et al.* (1987) reported on the appearance of the North American species *Paratenuisentis ambiguus*, a specialist parasite of the American eel *Anguilla rostrata*, in Atlantic eels *Anguilla anguilla* in the River Weser. This species had never previously been recorded outside the USA. Its presence in the River Weser was related to a previous introduction there of *Gammarus tigrinus*. This species was its only natural inter-mediate host in the USA, and it had been deliberately introduced into the River Weser some years previously to replace the indigenous gammarid fauna which had disappeared due to pollution. The amphipods came from a source in Britain, and so it is unlikely that the parasite was introduced with them. The acanthocephalan was able to transfer to the native European eel without any difficulty: it had no detrimental effect on its new host and was able to reproduce successfully within it. At that time (1986) 87.6% of the eels in the Weser were infected, with a mean of 98.2 parasites per eel, and 26.5% *G. tigrinus* were infected. By 1995, the parasite had spread to the River Rhine itself, where it completely dominated the helminth communities in eels with prevalence levels and abundances at two localities of 39.3% (16.4) and 56.7% (13.5) respectively. Other acanthocephalan species were present: *Acanthocephalus lucii* and *A. anguillae* at 1.6% in one locality and *Pomphorhynchus laevis* at 6.5% and 8.3% in the two localities, but their low levels reflected the dominance of the crustacean fauna by *G. tigrinus*.

By 1999 the situation had changed again. Sures & Streit (2001) sampled eels and crustaceans at two localities on the Rhine, near Karlsruhe and further north near Worms, and closer to the junction of the Maine–Danube canal through which several species of crustaceans had migrated into the Rhine system. Acanthocephalans still dominated the helminth fauna of eels at both sites, with *P. ambiguus* remaining the dominant species. Prevalence levels, however, were much lower at 14.3% near Worms and 36.8% at Alb (Table 9.3). By contrast, infection levels of the other three species had increased and especially at Alb. This appeared to be related to the low levels of *G. tigrinus* at Alb, where the crustaceans were now dominated by *Asellus aquaticus* and *Gammarus pulex*. At Worms, *G. tigrinus* survived but the crustacean fauna was dominated by *Corophium curvispinum*, an invader via the canal. The decline in *G. tigrinus* is due to the decrease in salinity in the river plus the effect of the species that have invaded through the canal. Thus, in this locality, the crustacean fauna has shown pronounced changes over a period of only four years and this has been a major influence on the

Table 9.3. *Abundance of acanthocephalans in eels* Anguilla anguilla *in two localities in the River Rhine*

	Worms		Alb	
	Prevalence (%)	Abundance Mean (± SD)	Prevalence (%)	Abundance Mean (± SD)
Paratenuisentis ambiguus	14.3	4.3 (18.7)	36.8	6.0 (17.2)
Acanthocephalus anguillae	2.9	0.1 (0.3)	15.8	0.6 (1.7)
Pomphorhynchus laevis	0	0	15.8	0.2 (0.4)

Source: Modified from Sures & Streit (2001).

acanthocephalan fauna. The introduction of *P. ambiguus* itself appears to have had no impact on the ecosystem: but the changes in abundance of the crustacean species and the arrival of introduced species have. Here, in this example, introduced crustaceans have had major effects on the native and introduced acanthocephalan species rather than vice versa.

The final example has been reported by Bauer *et al.* (2000) and is the converse of the situation in the River Rhine. They examined the infection levels of *P. laevis* and its effects on the behaviour of its intermediate host in two species of *Gammarus* in the River Ouche. The resident species was *G. pulex*, infected specimens of which showed the typical reversal of their normal negative phototactic behaviour to light when infected (Fig. 5.12). However, *G. roeseli*, a recent coloniser of the river which lived in sympatry with *G. pulex*, behaved normally, i.e. negatively phototacic in response to light even when infected with cystacanths. The parasite was able to infect *G. roeseli*, but was unable to alter its behaviour. The authors suggested that manipulation of gammarid behaviour was a feature of *P. laevis*'s adaptation to its resident host *G. pulex*, and resistant individuals of *G. roeseli* may have been selected during the invasion process. Since the effect of this was to attract selective predation on *G. pulex*, the failure of the parasite to alter the behaviour of *G. roeseli* conferred a selective advantage on an invading species, i.e. this could be an example of parasite-mediated competition.

In many respects, the invasion of the River Culm by *Pomphorhynchus laevis* supports their interpretation (pp. 98, 171). The parasite had only recently been introduced into the river with its definitive host, chub *Leuciscus cephalus*. The intermediate host *Gammarus pulex* has always been present in the river, but the population had been isolated from the rest of the metapopulation in the River Exe catchment by a zone of heavy pollution from paper mills on the lower reaches of the Culm, and a preliminary study has suggested that the population may be genetically distinct (Kennedy, unpublished). In effect, therefore, a different strain (ecologically equivalent to species) of *G. pulex* was encountering the invading *P. laevis* for the first time, setting up a new host–parasite system. The inability of the parasite to affect the behaviour of infected *G. pulex* (in contrast to Table 5.7) and of *G. pulex* to tolerate higher densities of *P. laevis* than is normal in the *P. laevis*–*G. pulex* system (Table 5.8) strongly sup\ports the interpretation of Bauer *et al.* (2000). These examples illustrate clearly the role that acanthocephalan parasites can play in shaping the structure of amphipod communities and so affecting the food web and the ecosystem organisation.

10

Conclusions and overview

It is suggested in Chapter 1 that although the phylum Acanthocephala is a small one with few species compared with the parasitic platyhelminths and nematodes and appears to show little diversity, nevertheless acanthocephalans are common, widespread and by any criteria they must be considered a successful group. It is also suggested that they exhibit a distinctive pattern of parasite–host co-evolution that differs from that found in other groups of parasites. As is the case with other groups of parasites, their anatomical simplicity is not evidence of degeneracy but rather of high levels of adaptation to their particular mode of life. Indeed, anatomical simplicity and similarity may bear little or no relationship to molecular diversity: morphological similarity and uniformity may in fact disguise molecular diversity. The evidence to date suggests strongly that much acanthocephalan diversity is to be found at the molecular level. Acanthocephalans may be understudied, but they should not be underestimated. They present a challenge to a number of paradigms: they show that anatomical and life cycle diversity is not important for the success of a group nor should the importance of a group be judged by the impact of its members on humans and/or their domestic animals. Acanthocephalans may have important and subtle ecological effects on ecosystems and food webs even though they do not exert pronounced pathological effects on their vertebrate hosts. More than any parasitic group, they are adapted to exploit whole ecosystems and not just their hosts in them.

Acanthocephalans have evolved a single basic pattern of organisation which has subsequently been fine-tuned in the course of evolution of different parasite–host systems. This is in marked contrast to the evolution of cestodes and digeneans which have, judged by morphological criteria, speciated to a far greater extent and have evolved complex and diverse life cycles and larval stages. In respect of fixed morphology and life cycles, but not of numbers of species, acanthocephalans are comparable to nematodes. The key aspects of the acanthocephalan plan are similarity in structure with little morphological variation and a fixed basic life cycle pattern of acanthor, cystacanth and adult with a single arthropod intermediate host – but with superimposed flexibility. Acanthocephalans are essentially a low-profile group: they have evolved to have virtually no impact on their vertebrate hosts, except when infrapopulation densities reach unusually high levels, but they manipulate their intermediate hosts to facilitate transmission in subtle but sophisticated, highly effective and adaptive ways. They keep their head below the vertebrate parapet, but above the invertebrate one. Their relationships with their intermediate hosts are a vital feature of the acanthocephalan co-evolutionary pattern and they are also a key to understanding the ecology and evolution of the group.

The acanthocephalans undoubtedly evolved as an aquatic group and the majority of species are still aquatic. Adults infect all groups of aquatic vertebrates with the exception of elasmobranchs. In addition to fish, this includes representatives of terrestrial groups that have retained or returned to an aquatic mode of life including chelonians, pinnipedes and cetaceans. They have also managed to adapt to a terrestrial mode of life, but not as successfully as, for example, the nematodes. Nematodes are able to infect vertebrates directly by free-living larvae penetrating their skin or indirectly by the ingestion of free-living larvae attached to vegetation, and so they are able to infect herbivores and ungulates in particular. Acanthocephalans have never evolved a method of infecting herbivores or of direct entry to grazers. This may be of little importance in water where herbivores and grazers are relatively uncommon, but it has limited their success on land. They still rely on ingestion for transmission but have never managed to lose their intermediate host as have nematodes nor to manipulate it in the way that some digeneans, notably *Dicrocoelium dendriticum*, have in order to infect grazing sheep (see Kennedy, 1975). They are thus rare parasites of grazing reptiles, marsupials and mammals; helminth communities of these are dominated by nematodes. Acanthocephalan reliance on food webs for transmission and ability to incorporate paratenic hosts into their life cycles results in their infecting consumers and predators, even top predators, but not herbivores.

In the Introduction, a number of questions were posed, the answers to which, it was hoped, might explain the success of the acanthocephalans. These questions can now be answered. The apparent limitations of the obligatory two-host life cycle can be overcome to a considerable degree. Eggs of terrestrial species can probably survive for much longer periods on land than can eggs of aquatic species in water. The life cycle can be extended facultatively by the incorporation of a paratenic host into it and by post-cyclic transmission, but it cannot be shortened either by loss of a host or by progenesis. Acanthocephalans are distributed widely on a global scale and have been reported from all biomes, but the majority of species are aquatic and are found in temperate latitudes. Distributions may become more patchy and discontinuous as the scale is reduced. Continental, post-glacial and regional patterns of distribution can nevertheless still be detected in some cases, but local conditions, especially the availability of intermediate hosts, are of vital importance as determinants of local patchiness.

Acanthocephalans have specialised in adding hosts laterally: they can show very wide specificity at either intermediate or definitive host level and the hosts are related by diet rather than by phylogengy. They may show a preference for one or several species, but they are able to use other species, which extends their range in time and space and enables them to survive in different localities or when a preferred host is temporarily or permanently absent from a locality. They can acquire new hosts by host capture and they can adapt to local conditions by formation of discrete strains which use different hosts in different localities. They have a minimal impact on their vertebrate hosts although they may cause local damage to the intestinal wall and, perhaps related to this and the cost-effectiveness of mounting a response, they are tolerated by these hosts. Both host and parasite appear to benefit from this arrangement and evolution has proceeded along the path of peaceful co-existence rather than as an arms race. Their impact on their intermediate hosts may, however, be severe, ranging from direct mortality through castration to reduced reproductive capacity in addition to the manipulation of colour and/or behaviour that is so effective in ensuring transmission and so precisely adaptive to the feeding behaviour of the preferred definitive host.

Acanthocephalans appear to be able to regulate their own population densities through intraspecific competition for space in their vertebrate hosts, yet also be able to respond to, and take advantage of, changes in abiotic conditions in a habitat and so build up a population rapidly. At normal levels of infection, the regulatory mechanisms may not operate but over the long term populations may be very stable, suggesting that density dependent regulation is effective. They do not appear to be able to regulate their intermediate host population, but regulation of the suprapopulation through competition in the vertebrate host should keep levels in the invertebrate host low enough to ensure that any deleterious effects on the arthopods do not lead to disappearance of that population. The balance is a fine one: if the impact is too severe the intermediate host population may decline below the threshold level for parasite persistence, but if the impact is not severe enough, transmission rates to the definitive host may be too low for parasite survival. Occasional breakdown of the regulatory mechanisms may lead to short-lived epizootics.

The role of acanthocephalans in helminth communities in hosts is variable. In rich and interactive communities they contribute to species richness but appear to make little, or occasionally only a minor, contribution to community structuring. Their impact is much greater in less diverse isolationist communities, and especially in fish hosts in northern and southern temperate regions where they may and often do dominate the helminth communities and contribute significantly to their structuring as well as richness. They are generally able to avoid interspecific competition by niche selection and overdispersion. However, if numbers of co-occurring competitor species increase, acanthocephalan species can compete effectively with other acanthocephalans as well as winning against members of other parasitic groups in the same guild.

Acanthocephalans disperse effectively, using their vertebrate hosts, both definitive and paratenic, to reach new localities. They can use preferred, additional and migratory hosts for the same purpose and as temporary refuges to avoid local extinctions. Local colonisations are common, but success depends to a high degree on the presence of the required hosts at the right densities and, of course, on chance. Local extinctions also appear to be fairly common, and populations may be maintained by rescue effects from a metapopulation. Extinction at higher levels, i.e. regional or global, is probably very rare as the ability to use a wide range of hosts and to adapt to the local conditions will ensure their survival.

The effects of acanthocephalans on ecosystems can be profound in view of their ability to manipulate their arthropod hosts. This can have a major impact on the structure of food webs and on predator–prey relationships and predator switching, and so on the flow of energy and materials through a food web. This can equally affect the population densities of other species at the same trophic levels. In other cases, acanthocephalans may affect the potential competitive interactions between an invading and resident species. In respect of direct interest to humans, their ability to take up heavy metals has the potential to make them very sensitive bioindicators: more sensitive than free-living species, they may also affect the uptake of metals by their hosts.

The success of the acanthocephalans as a group is not due to any single one of the above factors, but rather to the sum of them all. It is the combination of characteristics that enables them to overcome the apparent limitations of their life cycle and reduced ability to disperse (Table 10.1). Even narrow specificity does not of necessity limit a species. *Paratenuisentis ambiguus* is an eel specialist and in North America its preferred definitive host is the American eel

Table 10.1. *Summary of means by which acanthocephalans overcome or avoid their apparent life cycle limitations and related ecological problems*

Problem	Solution
Two-host life cycle	Add paratenic host and/or post-cyclic transmission
Limited dispersal ability	Use vertebrate hosts both preferred and secondary
Require two hosts to colonise	Wide specificity to one host and use of key species
Avoidance of extinction	Wide specificity and ability to adapt to new hosts
Dispersal in time	Eggs, cystacanths and paratenic stages are resting stages
Transmission to vertebrate	Manipulation of intermediate host behaviour
Overinfection of vertebrates	Regulation of infrapopulation density
Minimising competition	Host partitioning, niche segregation and competitive exclusion
Establishing new populations	Adaptation to local hosts, host capture and strain formation
Unfavourable times	Rescue effects and temporary use of less suitable hosts

Anguilla rostrata (Hoffman, 1999). This is a widespread and adaptable species occurring along the eastern seaboard, as is its intermediate host *Gammarus tigrinus*. When the parasite and the intermediate host species were translocated to Europe, the parasite was able to infect the European eel *Anguilla anguilla* and the parasite was able to dominate eel helminth communities in the River Rhine (Sures & Streit, 2001). The equivalent European species is *Acanthocephalus clavula*, for which *A. anguilla* is the preferred definitive host (Chubb, 1964). The European eel is itself widespread but the parasite can infect and mature in a wide range of non-anguillid hosts in addition. Specificity to the intermediate host may be fairly strict on a local scale: *Asellus meridianus* is its only intermediate host in the British Isles and is widespread and a keystone species in many communities. On continental Europe, however, the parasite can use other keystone species of intermediate hosts including *Echinogammarus pungens* in some localities (Dezfuli *et al.*, 1991b).

The use of a keystone species as an intermediate host is both a strength and limitation. Many species of vertebrate will feed on it and if the parasite shows wide specificity it will reproduce and survive in several of them. The trade-off is that the parasite will survive but not reproduce in other vertebrate species that also feed on the intermediate host, and these parasites are effectively dead to the population. Acanthocephalans end up where they are ingested but wide specificity does enable them to make the most of it. However, narrow specifciy may limit their local distribution to patches where the intermediate host is abundant. Taken over a longer period, acanthocephalans may adapt to local conditions and eventually form a distinct strain that is able to use, and has adapted to, different hosts in particular regions. Recent research using molecular methodology is suggesting that there may be far more variation at subspecific levels than had originally been envisaged. This may be evidenced by biological and distributional characteristics rather than by morphological ones and there may turn out to be many more sibling species and species complexes in the acanthocephalans than have so far been identified. Founder effects and isolation will assist in local genetic divergence and it is possible that much acanthocephalan–host co-evolution occurs at the local level and in relation to local conditions.

An excellent example of this is the colonisation of Ireland by the acanthocephalans of fish. The original post-glacial fish fauna of Ireland was exclusively salmonids and other migratory species such as eels. Cyprinids and fish of other families are later additions and were introduced by humans. In Ireland *Pomphorhynchus laevis* has adapted to use *Salmo trutta* as its preferred definitive host and *Gammarus duebeni* as its intermediate in the absence there of *Leuciscus cephalus* and *Barbus barbus*, its preferred definitive hosts, and *Gammarus pulex*, its preferred intermediate host in Britain; and has been recognised on both morphological and molecular grounds as belonging to a distinct strain (O'Mahony, 2004a,b). *Acanthocephalus anguillae* also uses *Leuciscus cephalus* as its preferred host in Britain and continental Europe, but in the absence of this species from Ireland it uses *Anguilla anguilla* as its normal host; *Asellus aquaticus*, its intermediate host, is present in Ireland. By contrast, *Echinorhynchus truttae* uses *Salmo trutta* as its normal definitive host in Ireland and elsewhere, but has adapted to *Gammarus duebeni* as intermediate host in Ireland.

The most unusual species in Ireland is *Neoechinorhynchus rutili*, which appears to be very uncommon (Holland & Kennedy, 1997) and has not been recorded there from *S. trutta*, even though it is common in this host in Scotland (Dorucu *et al.*, 1995) and though brown trout are widespread and common in Ireland. This reduction in host availability of some of the species in Ireland forces them to share hosts more than is normal in England (Lyndon & Kennedy, 2001) and so interspecific competition is commoner there (Byrne *et al.*, 2003).

Acanthocephalans are unspectacular but are very efficient just because they are so adaptable. There is a tendency to evaluate the importance of phyla and species in terms of their direct effect on humans. Acanthocephalans are often considered unimportant parasites, largely because they do not affect *Homo sapiens* (Ashford & Crewe, 1998, list records of only five species from humans, none of which reproduces in this host and all are considered accidental infections) or our domestic or farm animals, these latter all being herbivores. The lack of biodiversity in acanthocephalans makes them of less interest to many scientists. As with many parasites, complex life cycles attract some interest but the perspective may be incorrect: what may for us appear a difficult transmission or climb up a food chain by a parasite may not be so for the parasite. Yet, as has been demonstrated, it is the impact of acanthocephalans on the food web that makes them so important and their usefulness to us as bioindicators of heavy metals.

The adaptability that they demonstrate may be fascinating, but even this may only be a part of the pattern. We only see what is present in a gene pool when it is expressed: the potential remains hidden until then. The ability of acanthocephalans to switch hosts, set up new systems, develop new strains and survive when occasion demands is extraordinary and it is pertinent to wonder what remains unexpressed in the gene pool. To some degree it is possible to predict what species of acanthocephalan one might expect to find in a habitat if the potential intermediate and definitive hosts are known: for example, a shallow, productive lake with good populations of perch *Perca fluviatilis* and *Asellus aquaticus* might well be expected to be dominated by *Acanthocephalus lucii*, whereas it could be predicted a fast-flowing river on the European continent with good populations of barbel *Barbus barbus* and *Gammarus* sp. would be dominated by *Pomphorhynchus laevis*. General predictions are also possible: for example, that acanthocephalan richness will relate to the size of a water body (Kennedy, 1978b), to the nearness of this body to other localities (Poulin & Morand, 1999) and to the presence of other fish hosts in a locality (Leong & Holmes, 1981). However, it would have been far more difficult to predict that a species of *Onicola*, a genus whose members are normally found in felids and canids, would be able to infect birds, yet *O. pomastomi* appears to have travelled from Australia via birds to Malaysia where it infects bandicoots. As far as is known, no species of acanthocephalan has yet become globally extinct!

The theme running throughout this account has been that it is necessary to understand the ecology of the acanthocephalans, not their phylogeny, in order to understand their patterns of survival and success. Acanthocephalans adapt to flourish in the local ecosystems and conditions prevailing therein. Aquatic species use ecologically important keystone species as their intermediate hosts and the modified behaviour of these induced by the parasites facilitates transmission through the food web to their definitive hosts (Mouritsen & Poulin, 2002). The various suitable vertebrate definitive hosts infected are always ecologically, not phylogenetically, related (Poulin, 2005). Utilisation of paratenic hosts and post-cyclic transmission also relies on trophic links. The consequent lack of specialisation in many acanthocephalan species should not be viewed as a disadvantage but rather as an insurance: acanthocephalans on the whole have not followed the evolutionary path of narrow specialisation and host specificity so apparent in monogeneans and cestodes. Specificity to the intermediate host may appear to be narrower than that to the definitive host in the aquatic eoacanthocephalans, although not in the terrestrial archiacanthocephalans, but everything about the acantho-cephalans makes more sense in an ecological context than in a phylo-genetic one. More than any other group of parasites, they have adapted to take advantage of the whole food web and not just particular host species. In this they are unique and it is this more than anything else that distinguishes their parasite–host co-evolutionary patterns from those of other helminth parasite groups. As Anderson & May (1982) have suggested, there are many paths of co-evolution. Acanthocephalans have evolved towards peaceful co-existence with their vertebrate hosts, but aggression with their intermediate hosts. Both aggression and peaceful are relative terms and the different species of acanthocephalans lie at different points along their gradients.

Most of the examples given, the conclusions reached and the preceding generalisations refer primarily to aquatic, especially freshwater, species and to the temperate Holarctic. In this they largely reflect the information available, as more freshwater aquatic species have been studied in detail than marine species, and more aquatic ones than terrestrial ones. It is also the case, of course, that there are more aquatic species of acanthocephalans than terrestrial ones, but it is nevertheless pertinent to ask whether these generalisations do apply across the spectra of hosts, habitats and climatic zones. Obviously the identity of the hosts will change on a global and local scale from Alaska to Argentina, but the patterns may not. The opening remark in this book referred to MacArthur's search for patterns. What is now evident is that there are patterns discernible in the ecology of acanthocephalans regardless of continent and habitat. Patterns in distribution, relating to post-glacial and even earlier events have been demonstrated in the Americas and in Europe, as have colonisation events on Pacific islands and in Australasia. Wide specificity and the ability to use a wide range of hosts at either definitive or intermediate host level is evident in terrestrial, freshwater and marine species and the hosts are always ecologically related. The use of keystone arthropod species as intermediate hosts is common to terrestrial, freshwater and marine species. All species, regardless of host or habitat, appear to be capable of manipulating their intermediate hosts to facilitate transmission. No species has been shown to be consistently pathogenic to its definitive host(s). The ability of acanthocephalan species to regulate their own population levels by intraspecific competition for space has been demonstrated in species from mammals, birds and fish, as has the ability of acanthocephalan species to compete successfully with other species of acanthocephalans and with cestodes and nematodes.

Much more information on acanthocephalan biology and ecology is needed from marine and terrestrial species and from mammalian and bird hosts, and with more information it is likely that new patterns will emerge as well as more variations on the existing patterns. It may be, for example, that paratenic hosts are really more commonly employed by archiacanthocephalans and this in turn may relate to the greater difficulty of transmission on land. It may also emerge that, for similar reasons, terrestrial species in general show wider specificity to their intermediate hosts and narrower specificity to their definitive hosts than aquatic species. A study of patterns at the molecular level has scarcely begun, but this promises to be an exciting and informative field for future research. So far, the search for patterns in the ecology of the Acanthocephala has proved very productive and has provided an invaluable insight into the group. The patterns identified to date appear to be real ones and if so they are unlikely to be changed.

References

Abu-Madi, M. A., Behnke, J. M., Lewis, J. W. & Gilbert, F. S. (2000). Seasonal and site specific variation in the component community structure of intestinal helminths in *Apodemus sylvaticus* from three contrasting habitats in south-east England. *Journal of Helminthology*, **74**, 7–15.

Abu-Mahdi, M. A., Lewis, J. W., Mikhail, M., El-Nagger, M. E. & Behnke, J. M. (2001). Monospecific helminth and arthropod infections in an urban population of brown rats from Doha, Qatar. *Journal of Helminthology*, **75**, 313–20.

Abu-Mahdi, M. A., Behnke, J. M., Mikhail, M., Lewis, J. W. & Al-Kaabi, M. L. (2005). Parasite populations in the brown rat *Rattus norvegicus* from Doha, Qatar between years: the effect of host age, sex and density. *Journal of Helminthology*, **79**, 105–11.

Aho, J. M. (1990). Helminth communities of amphibians and reptiles: comparative approaches to understanding patterns and processes. In *Parasite Communities: Patterns and Processes*, ed. G. W. Esch, A. O. Bush, and J. M. Aho. London & New York: Chapman & Hall, pp. 156–95.

Aho, J. M., Bush, A. O. & Wolfe, R. W. (1991). Helminth parasites of bowfin (*Amia calva*) from South Carolina. *Journal of the Helminthological Society of Washington*, **58**, 171–5.

Aho, J. M., Mulvey, M., Jacobsen, K. C. & Esch, G. W. (1992). Genetic differentiation among congeneric acanthocephalans in the red-bellied slider turtle. *Journal of Parasitology*, **78**, 974–81.

Allan, J. C., Craig, P. S., Sherington, J. et al. (1999). Helminth parasites of the wild rabbit *Oryctolagus cuniculus* near Malham Tarn, Yorkshire, UK. *Journal of Helminthology*, **73**, 289–94.

Allely, Z., Moore, J. & Gotelli, N. J. (1992). *Moniliformis moniliformis* infection has no effect on some behaviours of the cockroach *Diploptera punctata*. *Journal of Parasitology*, **78**, 524–6.

Aloo, P. A. (2002). A comparative study of helminth parasites from the fish *Tilapia zilli* and *Oreochromis leucostictus* in Lake Naivasha and Oloidien Bay, Kenya. *Journal of Helminthology*, **76**, 95–102.

Amin, O. M. (1978). On the crustacean hosts of larval acanthocephalan and cestode parasites in south-western Lake Michigan, U.S.A. *Journal of Parasitology*, **64**, 842–5.

Amin, O. M. (1984). Interspecific variability in the genus *Acanthocephalus* (Acanthocephala: Echinorhynchidae) from North American freshwater fishes, with a key to species. *Proceedings of the Helminthological Society of Washington*, **51**, 238–40.

Amin, O. M. (1985a). Classification. In *Biology of the Acanthocephala*, ed. D. W. T. Crompton & B. B. Nickol. Cambridge: Cambridge University Press, pp. 27–72.

Amin, O. M. (1985b). Hosts and geographic distribution of *Acanthocephalus* (Acanthocephala: Echinorhynchidae) from North American freshwater fishes, with a discussion of species relationships. *Proceedings of the Helminthological Society of Washington*, **52**, 210–20.

Amin, O. M. (1986). On the species and populations of the genus *Acanthocephalus* (Acanthocephala: Echinorhynchidae) from North American freshwater fishes: a cladistic analysis. *Proceedings of the Biological Society of Washington*, **99**, 574–9.

Amin, O. M. (1987). Acanthocephala from lake fishes in Wisconsin: ecology and host relationships of *Pomphorhynchus bulbocolli* (Pomphorhynchidae). *Journal of Parasitology*, **73**, 278–89.

Amin, O. M. & Burrows, J. M. (1977). Host and seasonal associations of *Echinorhynchus salmonis* (Acanthocephala: Echinorhynchidae) in Lake Michigan fishes. *Journal of the Fisheries Research Board of Canada*, **34**, 325–31.

Amin, O., Burns, L. A. & Redlin, M. J. (1980). The ecology of *Acanthocephalus parksidei* (Acanthocephala: Echinorhynchidae) in its isopod intermediate host. *Proceedings of the Helminthological Society of Washington*, **47**, 124–31.

Anderson, R. M. (1982). Epidemiology. In *Modern Parasitology*, ed. F. E. G. Cox, 2nd edn. Oxford: Blackwell Scientific, pp. 204–51.

Anderson, R. M. & Gordon, D. M. (1982). Processes influencing the distribution of parasites numbers within host populations with special emphasis on parasite-induced host mortalities. *Parasitology*, **85**, 373–98.

Anderson, R. M. & May, R. M. (1978). Regulation and stability of host-parasite population interactions: I. Regulatory processes. *Journal of Animal Ecology*, **47**, 219–47.

Anderson, R. M. & May, R. M. (1979). Population biology of infectious diseases: 1. *Nature, London*, **280**, 361–7.

Anderson, R. M. & May, R. M. (1982). Coevolution of hosts and parasites. *Parasitology*, **85**, 411–26.

Andreassen, J. (1975a). Immunity to the acanthocephalan *Moniliformis dubius* infections in rats. *Norwegian Journal of Zoology*, **23**, 195–6.

Andreassen, J. (1975b). Reaginic antibodies in response to *Moniliformis dubius* infections in rats. *Norwegian Journal of Zoology*, **23**, 196.

Andrewartha, H. A. & Birch, L. C. (1954). *Distribution and Abundance of Animals*. Chicago: University of Chicago Press.

Arai, H. P. (1989). Acanthocephala. In *Guide to the Parasites of Fishes of Canada*, Part III, ed. L. Margolis and Z. Kabata, Canadian Special Publications of Fisheries and Aquatic Sciences, pp. 1–90.

Arnold, S. E. & Crompton, D. W. T. (1987). Survival of shelled acanthors of *Moniliformis moniliformis* under laboratory conditions. *Journal of Helminthology*, **61**, 306–10.

Arthur, J. R. & Albert, E. (1994). A survey of the parasites of Greenland halibut (*Reinhardtius hippoglossoides*) caught off Atlantic Canada, with notes on their zoogeography in the fish. *Canadian Journal of Zoology*, **72**, 765–78.

Ashford, R. W. & Crewe, W. (1998). *The Parasites of Homo sapiens*. Liverpool: Liverpool School of Tropical Medicine.

Ashley, D. C. & Nickol, B. B. (1989). Dynamics of the *Leptorhynchoides thecatus* (Acanthocephala) suprapopulation in a Great Plains reservoir. *Journal of Parasitology*, **75**, 46–54.

Avery, R. A. (1971). Helminth populations in newts and their tadpoles. *Freshwater Biology*, **1**, 113–9.

Awachie, J. B. E. (1965). The ecology of *Echinorhynchus truttae* Schrank, 1788 (Acanthocephala) in a trout stream in North Wales. *Parasitology*, **55**, 747–62.

Awachie, J. B. E. (1966). The development and life history of *Echinorhynchus truttae* Schrank, 1788 (Acanthocephala). *Journal of Helminthology*, **40**, 11–32.

Awachie, J. B. E. (1967). Experimental studies on some host-parasite relationships of the Acanthocephala. Co-invasion of *Gammarus pulex* L. by *Echinorhynchus truttae* Schrank, 1788 and *Polymorphus minutus* (Goeze, 1782). *Acta Parasitologica Polonica*, **15**, 69–74.

Awachie, J. B. E. (1972). Experimental studies on some host-parasite relationships of the Acanthocephala. Effects of primary heavy infection and superimposed infection of *Salmo trutta* L. by *Echinorhynchus truttae* Schrank, 1788. *Acta Parasitologica Polonica*, **20**, 375–82.

Aznar, F. J., Balbuena, J. A. & Raga, J. A. (1994). Helminth communities of *Pontoporia blainvillei* (Cetacea: Pontoporiidae) in Argentine waters. *Canadian Journal of Zoology*, **72**, 702–6.

Aznar, F. J., Badillo, F. J. & Raga, J. A. (1998). Gastrointestinal helminths of loggerhead turtles (*Caretta caretta*) from the western Mediterranean: constraints on community structure. *Journal of Parasitology*, **84**, 474–9.

Bafundo, K. W., Wilhelm, W. E. & Kennedy, M. L. (1980). Geographical variation in helminth parasites from the digestive tract of Tennessee racoons, *Procyon lota*. *Journal of Parasitology*, **66**, 134–9.

Bakker, T. C. M., Mazzi, D. & Zola, S. (1997). Parasite induced changes in behaviour and colour make *Gammarus pulex* more prone to fish predators. *Ecology*, **78**, 1098–104.

Balbuena, J. A. & Raga, J. A. (1993). Intestinal helminth community of the long-finned pilot whale (*Globicephala melas*) off the Faroe Islands. *Parasitology*, **106**, 327–33.

Balbuena, J. A. & Raga, J. A. (1994). Intestinal helminths as indicators of segregation and social structure of pods of long-finned pilot whales (*Globicephala melas*) off the Faroe Islands. *Canadian Journal of Zoology*, **72**, 443–8.

Baldanova, D. R. & Pronin, N. M. (2001). *Acanthocephalans (Phylum Acanthocephala) of Baikal*. Novosbirsk: Nauka.

Bangham, R. V. (1955). Studies on fish parasites of Lake Huron and Manitoulin Island. *American Midland Naturalist*, **53**, 184–94.

Barger, M. A. & Nickol, B. B. (1998). Structure of *Leptorhynchoides thecatus* and *Pomphorhynchus bulbocolli* (Acanthocephala) eggs in habitat partitioning and transmission. *Journal of Parasitology*, **84**, 534–7.

Barger, M. A. & Nickol, B. B. (1999). Effects of coinfection with *Pomphorhynchus bulbocolli* on development of *Leptorhynchoides thecatus* (Acanthocephala) in amphipods (*Hyalella azteca*). *Journal of Parasitology*, **85**, 60–3.

Barnard, C. J., Kulis, K., Behnke, J. M., *et al.* (2003). Local variation in helminth burdens of bank voles (*Clethrionomys glareolus*) from ecologically similar sites: temporal stability and relationships with hormone concentrations and social behaviour. *Journal of Helminthology*, **77**, 185–95.

Barnes, R. D. (1963). *Invertebrate Zoology*. Philadelphia: W. B. Saunders.

Barnes, R. S. K., Calow, P. & Olive, P. T. W. (1988). *The Invertebrates: A New Synthesis*. Oxford: Blackwell Scientific.

Bartoli, P., Morand, S., Riutort, J-J. & Combes, C. (2000). Acquisition of parasites correlated with social rank and behavioural changes in a fish species. *Journal of Helminthology*, **74**, 289–93.

Bates, R. M. & Kennedy, C. R. (1990). Interactions between the acanthocephalans *Pomphorhynchus laevis* and *Acanthocephalus anguillae* in rainbow trout: testing an exclusion hypothesis. *Parasitology*, **100**, 435–44.

Bates, R. M. & Kennedy, C. R. (1991a). Site availability and density-dependent constraints on the acanthocephalan *Pomphorhynchus laevis* in rainbow trout *Oncorhynchus mykiss* (Walbaum). *Parasitology*, **102**, 405–10.

Bates, R. M. & Kennedy, C. R. (1991b). Potential interactions between *Acanthocephalus anguillae* and *Pomphorhynchus laevis* in their natural hosts chub, *Leuciscus cephalus* and the European eel, *Anguilla anguilla. Parasitology*, **102**, 289–97.

Bauer, A., Trouve, S., Gregoire, A., Bollache, L. & Cezilly, F. (2000). Differential influence of *Pomphorhynchus laevis* (Acanthocephala) on the behaviour of native and invader gammarid species. *International Journal for Parasitology*, **30**, 1453–7.

Behnke, J. M., Barnard, C. J., Mason, N. *et al.* (2000). Intestinal helminths of spiny mice (*Acomys cahirinus dimidiatus*) from St Katherine's Protectorate in the Sinai, Egypt. *Journal of Helminthology*, **74**, 31–43.

Behnke, J. M., Barnard, C. J., Bajer, A. *et al.* (2001). Variation in the helminth community structure in bank voles (*Clethrionomys glareolus*) from three comparable localities in the Mazury Lake District region of Poland. *Journal of Helminthology*, **123**, 401–14.

Bell, G. & Burt, A. (1991). The comparative biology of parasite species diversity: internal helminths of freshwater fish. *Journal of Animal Ecology*, **60**, 1047–64.

Bertocchi, D. & Francalanci, G. (1963). Grave infestazione da *Echinorhynchus truttae* Schrank in trote iride di allevamento (*Salmo gairdneri*). *Veterinaria Italiana*, **14**, 475–81.

Bethel, W. M. & Holmes, J. C. (1973). Altered evasive behaviour and responses to light in amphipods harbouring acanthocephalan cystacanths. *Journal of Parasitology*, **59**, 945–56.

Bethel, W. M. & Holmes, J. C. (1974). Correlation of development of altered evasive behaviour in *Gammarus lacustris* (Amphipoda) harbouring cystacanths of *Polymorphus paradoxus* (Acanthocephala) with the infectivity to the definitive host. *Journal of Parasitology*, **60**, 272–4.

Bethel, W. M. & Holmes, J. C. (1977). Increased vulnerability of amphipods to predation owing to altered behaviour induced by larval acanthocephalans. *Canadian Journal of Zoology*, **55**, 110–15.

Beveridge, I. & Spratt, D. M. (1996). The helminth fauna of Australian marsupials: origins and evolutionary biology. *Advances in Parasitology*, **37**, 136–254.

Biserkov, V. & Kostadinova, A. (1998). Intestinal helminth communities in the green lizard, *Lacerta viridis*, from Bulgaria. *Journal of Helminthology*, **72**, 267–71.

Blaylock, R. B., Holmes, J. C. & Margolis, L. (1998). The parasites of Pacific halibut (*Hippoglossus stenolepis*) in the eastern North Pacific: host level influences. *Canadian Journal of Zoology*, **71**, 536–47.

Boes, J., Willingham, A. L. III., Shi Fuhui, Hu Xuguang, Eriksen, L., Nansen, P. & Stewart, T. B. (2000). Prevalence and distribution of pig helminths in the Dongting Lake Region (Hunan Province) of the People's Republic of China. *Journal of Helminthology* **74**, 45–52.

Boquimpani-Freitas, L., Vrcibradic, D., Vicente, J. J., Bursey, C. R., Rocha, C. F. D. & Van Sluys, M. (2001). Helminths of the horned leaf frog, *Proceratophrys appendiculata*, from southeastern Brazil. *Journal of Helminthology*, **75**, 233−6.

Borgsteede, F. H. M., Haenen, O. L. M., De Bree, J. & Lisitina, O. I. (1999). Parasitic infections of European eel (*Anguilla anguilla* L.) in the Netherlands. *Helminthologia*, **36**, 251−60.

Bozkov, D. K. (1980). Experimental studies on the passage of mature helminths from *Rana ridibunda* Pall. to *Bufo viridis* Laur. *Helminthologia*, **10**, 24−8.

Brattey, J. (1983). The effects of larval *Acanthocephalus lucii* on the pigmentation, reproduction and susceptibility to predation of the isopod *Asellus aquaticus*. *Journal of Parasitology*, **69**, 1172−3.

Brattey, J. (1986). Life history and population biology of larval *Acanthocephalus lucii* (Acanthocephala: Echinorhynchidae) in the isopod *Asellus aquaticus*. *Journal of Parasitology*, **72**, 633−45.

Brattey, J. (1988). Life history and population biology of adult *Acanthocephalus lucii* (Acanthocephala: Echinorhynchidae). *Journal of Parasitology*, **74**, 72−80.

Brown, A. F. (1986). Evidence for density-dependent establishment and survival of *Pomphorhynchus laevis* (Muller, 1776) (Acanthocephala) in laboratory-infected *Salmo gairdneri* Richardson and its bearing on wild populations in *Leuciscus cephalus* (L.). *Journal of Fish Biology*, **28**, 659−69.

Brown, A. F. (1987). Anatomical variability and secondary sexual characteristics in *Pomphorhynchus laevis* (Muller, 1776) (Acanthocephala). *Systematic Parasitology*, **9**, 213−19.

Brown, A. F. (1989). Seasonal dynamics of the acanthocephalan *Pomphorhynchus laevis* (Muller, 1776) in its intermediate and preferred definitive hosts. *Journal of Fish Biology*, **34**, 183−94.

Brown, A. F. & Pascoe, D. (1989). Parasitism and host sensitivity to cadmium: an acanthocephalan infection of the freshwater amphipod *Gammarus pulex*. *Journal of Applied Ecology*, **26**, 473−87.

Brown, A. F. & Thompson, D. B. A. (1986). Parasite manipulation of host behaviour: acanthocephalans and shrimps in the laboratory. *Journal of Biological Education*, **20**, 121−7.

Buckner, R. L. & Nickol, B. B. (1979). Geographic and host-related variation amongst species of *Fessisentis* (Acanthocephala) and confirmation of the *Fessisentis fessus* life cycle. *Journal of Parasitology*, **65**, 161−6.

Bullock, W. L. (1963). Intestinal histology of some salmonid fishes with particular reference to the histopathology of acanthocephalan infections. *Journal of Morphology*, **112**, 23−44.

Burlingame, P. L. & Chandler, A. C. (1941). Host-parasite relations of *Moniliformis dubius* (Acanthocephala) in albino rats, and the environmental nature of resistance to single and superimposed infections with this parasite. *American Journal of Hygiene*, **33**, 1−21.

Buron, I. de (1988). *Hypoechinorhynchus thermaceri* N.sp. (Acanthocephala: Hypoechinorhynchidae) from the deep sea zoarcid fish *Thermaceres andersoni* Rosenblatt and Cohen, 1986. *Journal of Parasitology*, **74**, 339−42.

Buron, I. de & Maillard, C. (1987). Transfert expérimental d'helminthes adultes chez les poissons par ichtiophagie et cannibalisme. *Annales de Parasitologie*, **62**, 188−91.

Buron, I. de & Morand, S. (2004). Deep-sea hydrothermal vent parasites: why do we not find more? *Parasitology*, **128**, 1−7.

Buron, I. de, Rebaud, F. & Euzet, L. (1986). Speciation and specificity of acanthocephalans. Genetic and morphological studies of *Acanthocephalus geneticus* sp.nov. parasitising *Arnoglossus laterna* (Bothidae) from the Mediterranean littoral (Sete-France). *Parasitology*, **92**, 165–71.

Bursey, C. R. & Goldberg, S. R. (1998). Helminths of the Canadian toad, *Bufo hemiophrys* (Amphibia: Anura) from Alberta, Canada. *Journal of Parasitology*, **84**, 617–18.

Buscher, H. N. (1965). Dynamics of the intestinal helminth fauna in three species of ducks. *Journal of Wildlife Management*, **29**, 772–81.

Bush, A. O. (1990). Helminth communities in avian hosts: determinants of pattern. In *Parasite Communities: Patterns and Processes*, ed. G. W. Esch, A. O. Bush and J. M. Aho, London & New York: Chapman & Hall, pp. 197–232.

Bush, A. O. & Holmes, J. C. (1986a) Intestinal helminths of lesser scaup ducks: patterns of association. *Canadian Journal of Zoology*, **64**, 132–41.

Bush, A. O. & Holmes, J. C. (1986b). Intestinal helminths of lesser scaup ducks: an interactive community. *Canadian Journal of Zoology*, **64**, 142–52.

Bush, A. O. & Kennedy, C. R. (1994). Host fragmentation and helminth parasites: hedging your bets against extinction. *International Journal for Parasitology*, **24**, 1333–43.

Bush, A. O., Aho, J. M. & Kennedy, C. R. (1990). Ecological versus phylogenetic determinants of helminth parasite community richness. *Evolutionary Ecology*, **4**, 1–20.

Bush, A. O., Heard, R. W. & Overstreet, R. M. (1993). Intermediate hosts as source communities. *Canadian Journal of Zoology*, **71**, 1358–63.

Bush, A. O., Lafferty, K. D., Lotz, J. M. & Shostack, A. W. (1997). Parasitology meets ecology on its own terms: Margolis et al. revisited. *Journal of Parasitology*, **83**, 575–83.

Bush, A. O., Fernandez, J. C., Esch, G. W. & Seed, J. R. (2001). *Parasitism: The Diversity and Ecology of Animal Parasites*. Cambridge: Cambridge University Press.

Bussche, R. A., Van den, Kennedy, M. L. & Wilhelm, W. E. (1987). Helminth parasites of the coyote (*Canis latrans*) in Tennesee. *Journal of Parasitology*, **73**, 327–32.

Byrne, C. J., Grey, C., Holland, C. & Poole, R. (2000). Parasite community similarity between four Irish lakes. *Journal of Helminthology*, **74**, 301–5.

Byrne, C. J., Holland, C. V., Poole, R. & Kennedy, C. R. (2002). Comparison of the macroparasite communities of wild and stocked brown trout (*Salmo trutta* L.) in the west of Ireland. *Parasitology*, **124**, 435–545.

Byrne, C. J., Holland, C. V., Kennedy, C. R. & Poole, W. R. (2003). Interspecific interactions between Acanthocephala in the intestine of brown trout: are they more common in Ireland? *Parasitology*, **127**, 399–409.

Byrne, C. J., Holland, C. V., Walsh, E., Mulligan, C., Kennedy, C. R. & Poole, W. R. (2004). Utilization of brown trout *Salmo trutta* by *Acanthocephalus clavula* in an Irish lake: is this evidence of a host shift? *Journal of Helminthology*, **78**, 201–6.

Callaghan, R. & McCarthy, T. K. (1996). Metazoan parasite assemblages of eels in the Dunkelin Catchment, Western Ireland. *Archives of Polish Fisheries*, **4**, 147–74.

Calvete, C., Blanco-Aguiar, J. A., Virgops, E., Cabezas-Diaz, S. & Villafuerte, R. (2004). Spatial variation in helminth community structure in the red-legged partridge (*Alectoris rufa* L.): effects of definitive host density. *Parasitology*, **129**, 101–13.

Camp, J. W. & Huizinga, H. W. (1979). Altered colour, behaviour and predation susceptibility of the isopod, *Asellus intermedius*, infected with *Acanthocephalus dirus*. *Journal of Parasitology*, **65**, 667–9.

Camp, J. W. & Huizinga, H. W. (1980). Seasonal population interactions of *Acanthocephalus dirus* (Van Cleave, 1931) in the creek chub, *Semotilus atromaculatus*, and isopod, *Asellus intermedius*. *Journal of Parasitology*, **66**, 299—304.

Campbell, N. A. (1996). *Biology*, 4th edn. Menlo Park, California: Benjamin/ Cummings Publishing.

Carmichael, L. M. & Moore, J. (1991). A comparison of behavioural alterations in the brown cockroach *Periplaneta brunea*, and the American cockroach, *Periplaneta americana*, infected with the acanthocephalan, *Moniliformis moniliformis*. *Journal of Parasitology*, **77**, 931—6.

Carney, J. P. & Dick, T. A. (2000). Helminth communities of yellow perch (*Perca flavescens* (Mitchill)): determinants of pattern. *Canadian Journal of Zoology*, **78**, 538—55.

Cezilly, F., Gregoire, A. & Bertin, A. (2000). Conflict between co-occurring manipulative parasites? An experimental study of the joint influences of two acanthocephalan parasites on the behaviour of *Gammarus pulex*. *Parasitology*, **120**, 625—30.

Chapman, M. R., French, D. D. & Klei, T. R. (2002). Gastrointestinal helminths of ponies in Louisiana; a comparison of species currently prevalent with those present 20 years ago. *Journal of Parasitology*, **88**, 1130—4.

Chappell, L. H. (1969). Competitive exclusion between two intestinal parasites of the three-spined stickleback, *Gasterosteus aculeatus* L. *Journal of Parasitology*, **55**, 775—8.

Choudhury, A. & Dick, T. A. (2000). Richness and diversity of helminth communities in tropical freshwater fishes: empirical evidence. *Journal of Biogeography*, **27**, 935—56.

Choudhury, A., Hoffnagke, T. L. & Cole, R. A. (2004). Parasites of native and non-native fishes of the Little Colorado River, Grand Canyon, Arizona. *Journal of Parasitology*, **90**, 1042—53.

Chubb, J. C. (1964). Occurrence of *Echinorhynchus clavula* (Dujardin, 1845) nec Hamann, 1892 (Acanthocephala) in the fish of Llyn Tegid (Bala Lake), Merionethshire. *Journal of Parasitology*, **50**, 52—9.

Chubb, J. C. (1967). Host specificity of some Acanthocephala of freshwater fishes. *Helminthologia*, **8**, 1—4.

Chubb, J. C. (1982). Seasonal occurrence of helminths in freshwater fishes: IV. Adult Cestoda, Nematoda and Acanthocephala. *Advances in Parasitology*, **20**, 1—293.

Chubb, J. C. (2004). *Echinorhynchus salmonis* Muller, 1784 absent in Britain and Ireland: re-identification of museum specimens. *Folia Parasitologica*, **51**, 72—4.

Combes, C. (1991). Evolution of parasite life cycles. In *Parasite Host Associations: Coexistence or Conflict?*, ed. C. A. Toft, A. Aeschlimann & L. Bolis. Oxford: Oxford University Press, pp. 62—82.

Combes, C. (1996). Parasites, biodiversity and ecosystem stability. *Biodiversity and Conservation*, **5**, 953—62.

Combes, C. (2001). *Parasitism: The Ecology and Evolution of Intimate Interactions*. Chicago: University of Chicago Press.

Comiskey, P. & Mackenzie, K. (2000). *Corynosoma* spp. may be useful biological tags for saithe in the northern North Sea. *Journal of Fish Biology*, **57**, 525—8.

Cone, D. K., Marcogliese, D. J. & Watt, W. D. (1993). Metazoan parasite communities of yellow eels (*Anguilla rostrata*) in acidic and limed rivers of Nova Scotia. *Canadian Journal of Zoology*, **71**, 177—84.

Conneely, J. J. & McCarthy, T. K. (1984). The metazoan parasites of freshwater fishes in the Corrib catchment area, Ireland. *Journal of Fish Biology*, **24**, 363—75.

Conneely, J. J. & McCarthy, T. K. (1986). Ecological factors influencing the composition of the parasite fauna of the European eel, *Anguilla anguilla* (L.), in Ireland. *Journal of Fish Biology*, **28**, 207–19.

Connell, R. & Corner, A. H. (1957). *Polymorphus paradoxus* sp. nov. (Acanthocephala) parasitizing beavers and muskrats in Alberta, Canada. *Canadian Journal of Zoology*, **35**, 525–33.

Conway-Morris, S. & Crompton, D. W. T. (1982). The origins and evolution of the Acanthocephala. *Biological Reviews*, **57**, 85–115.

Coombs, I. & Crompton, D. W. T. (1991). *A Guide to Human Helminths*. London: Taylor & Francis.

Cornell, H. V. & Lawton, J. H. (1992). Species interactions, local and regional processes and limits to the richness of ecological communities: a theoretical perspective. *Journal of Animal Ecology*, **61**, 1–12.

Costa, G., Chubb, J. C. & Veltkamp, C. J. (2002). Cystacanths of *Bolbosoma vasculosum* in the black scabbard fish *Aphanopus carbo*, oceanic horse mackerel *Trachurus picturatus* and common dolphin *Delphinus delphis* from Madeira, Portugal. *Journal of Helminthology*, **74**, 113–20.

Cox F. E. G. (ed.) (1993). *Modern Parasitology: A Textbook of Parasitology*, 2nd edn. Oxford: Blackwell Scientific Publications.

Craig, H. L. & Craig, P. S. (2005). Helminth parasites of wolves (*Canis lupus*): a species list and an analysis of published prevalence studies in Nearctic and Palaearctic populations. *Journal of Helminthology*, **79**, 95–103.

Cribb, T. H., Anderson, G. R. & Dove, A. D. M. (2000). *Pomphorhynchus heronensis* and restricted movement of *Lutjanus carponotatus* on the Great Barrier Reef. *Journal of Helminthology*, **74**, 53–6.

Crofton, H. D. (1971a). A quantitative approach to parasitism. *Parasitology*, **62**, 179–93.

Crofton, H. D. (1971b). A model of host parasite relationships. *Parasitology*, **63**, 343–64.

Crompton, D. W. T. (1967). Studies on the haemocytic reaction of *Gammarus* spp., and its relationship to *Polymorphus minutus* (Acanthocephala). *Parasitology*, **57**, 389–401.

Crompton, D. W. T. (1970). *An Ecological Approach to Acanthocephalan Physiology*. Cambridge: Cambridge University Press.

Crompton, D. W. T. (1973). The sites occupied by some parasitic helminths in the alimentary tract of vertebrates. *Biological Reviews*, **48**, 27–83.

Crompton, D. W. T. (1985). Reproduction. In *Biology of the Acanthocephala*, ed. D. W. T. Crompton & B. B. Nickol, Cambridge: Cambridge University Press, pp. 213–72.

Crompton, D. W. T. & Nickol B. B. (eds.) (1985). *Biology of the Acanthocephala*. Cambridge: Cambridge University Press.

Crompton, D. W. T. & Walters, D. E. (1972). An analysis of the course of infection of *Moniliformis dubius* (Acanthocephala) in rats. *Parasitology*, **64**, 517–23.

Crompton, D. W. T. & Whitfield, P. J. (1968). The course of infection and egg production of *Polymorphus minutus* (Acanthocephala) in domestic ducks. *Parasitology*, **58**, 231–46.

Crompton, D. W. T., Arnold, S. & Barnard, D. (1972). The patent period and production of eggs of *Moniliformis dubius* (Acanthocephala) in the small intestine of male rats. *International Journal for Parasitology*, **2**, 319–26.

Crompton, D. W. T., Keymer, A. E., Walters, D. E., Arnold, S. E. & Marrs, R. W. (1988). Factors influencing the fecundity of *Moniliformis moniliformis* (Acanthocephala): constant dose and varied diet. *Journal of Zoology, London*, **214**, 221–34.

Curtis, M. (1979). Metazoan parasites of resident arctic char (*Salvelinus alpinus*) from a small lake on Southern Baffin Island. *Le Naturaliste Canadien*, **106**, 337–8.

Daniels, C. B. (1985). The effect of infections by a parasitic worm on swimming and diving in the water skink *Sphaeromorphus quoyii*. *Journal of Herpetology*, **19**, 160–2.

De Giusti, D. L. (1949). The life cycle of *Leptorhynchoides thecatus* (Linton), an acanthocephalan of fish. *Journal of Parasitology*, **35**, 437–60.

De Liberato, C., Berrilli, F., Di Cave, D., Russo, R., Tancioni, I. & Kennedy, C. R. (2002). Intestinal parasites of Italian barbel, *Barbus tyberinus* (Cypriniformes: Cyprinidae), from the River Tiber and first report of *Acanthocephalus clavula* (Acanthocephala) in the genus *Barbus*. *Folia Parasitologica*, **49**, 246–8.

DeMont, D. J. & Corkum, K. C. (1982). The life cycle of *Octospiniferoides chandleri* Bullock, 1957 (Acanthocephala: Neoechinorhynchidae) with some observations on parasite-induced, photophilic behavior in ostracods. *Journal of Parasitology*, **68**, 125–30.

Denny, M. (1969). Life cycles of helminth parasites using *Gammarus lacustris* as an intermediate host in a Canadian lake. *Parasitology*, **59**, 795–827.

Dezfuli, B. S., Zanini, N., Reggiani, G. & Rossi, R. (1991a). *Echinogammarus stammeri* (Amphipoda) as an intermediate host for *Pomphorhynchus laevis* (Acanthocephala) parasite of fishes from the River Brenta. *Boletino Zoologica*, **58**, 267–71.

Dezfuli, B. S., Rossetti, E., Zanini, N. & Rossi, R. (1991b). Occurrence of *Acanthocephalus clavula* (Acanthocephala) in *Echinogammarus pungens* (Amphipoda) from the River Adige. *Parassitologia*, **33**, 121–6.

Dezfuli, B. S., Rossetti, E., Fano, E. A. & Rossi, R. (1994). Occurrence of larval *Acanthocephalus anguillae* (Acanthocephala) in the *Asellus aquaticus* (Crustacea: Isopoda) from the River Brenta. *Bollettino di Zoologia*, **61**, 77–81.

Dezfuli, B. S., Rossetti, E., Bellettato, C. M. & Maynard, B. J. (1999). *Pomphorhynchus laevis* in its intermediate host *Echinogammarus stammeri* in the River Brenta, Italy. *Journal of Helminthology*, **73**, 95–102.

Dezfuli, B. S., Giari, L. & Poulin, R. (2000). Species associations among larval helminths in an amphipod intermediate host. *International Journal for Parasitology*, **30**, 1143–6.

Dezfuli, B. S., Giari, L., De Biaggi, S. & Poulin, R. (2001a). Associations and interactions among intestinal helminths of the brown trout, *Salmo trutta*, in Northern Italy. *Journal of Helminthology*, **75**, 331–6.

Dezfuli, B. S., Giarai, L. & Poulin, R. (2001b). Costs of intraspecific and interspecific host sharing in acanthocephalan cystacanths. *Parasitology*, **122**, 483–9.

Dezfuli, B. S., Volponi, S., Beltrami, I. & Poulin, R. (2002). Intra- and inter-specific density-dependent effects on growth in helminth parasites of the cormorant *Phalacrocorax carbo sinensis*. *Parasitology*, **124**, 537–44.

Dezfuli, B. S., Maynard, B. J. & Wellnitz, T. (2003). Activity levels and predator detection by amphipods infected with an acanthocephalan parasite, *Pomphorhynchus laevis*. *Folia Parasitologica*, **50**, 129–34.

Diamant, A. (1989). Ecology of the acanthocephalan *Sclerocollum rubrimaris* Schmidt and Paperna, 1978 (Rhadinorhynchidae: Gorgorhynchidae) from wild populations of rabbit fish (genus *Siganus*) in the northern Red Sea. *Journal of Fish Biology*, **34**, 387–97.

Di Cave, D., Kennedy, C. R., D'Amelio, S., Berrilli, F. & Orecchia, P. (1997). Structure of the metazoan parasite community of *Liza ramada* (Risso, 1826) in the Burano Lake. *Biologia Maritima Mediterrania*, **4**, 321–3.

Di Cave, D., Berrilli, F., De Liberato, C., Orecchia, P. & Kennedy, C. R. (2001). Helminth communities in eels from Adriatic coastal lagoons in Italy. *Journal of Helminthology*, **75**, 7–13.

Dobson, A. P. (1985). The population dynamics of competition between parasites. *Parasitology*, **66**, 317–47.

Dobson, A. P. (1986). Inequalities in the individual reproductive success of parasites. *Parasitology*, **92**, 675–82.

Dobson, A. P. (1988). Population biology of parasite induced changes in host behaviour. *Quarterly Reviews in Biology*, **63**, 139–65.

Dobson, A. P. (1990). Models for multi-species parasite-host communities. In *Parasite Communities: Patterns and Processes*, ed. G. W. Esch, A. O. Bush and J. M. Aho. London: Chapman & Hall, pp. 261–88.

Dobson, A. P. & Keymer, A. E. (1985). Life history models. In *Biology of the Acanthocephala*, ed. D. W. T. Crompton and B. B. Nickol, Cambridge: Cambridge University Press, pp. 347–84.

Dobson, A. P. & Roberts, M. (1994). The population dynamics of parasitic helminth communities. *Parasitology*, **109**, S97–S108.

Dogiel, V. A. (1964). *General Parasitology*. Edinburgh & London: Oliver & Boyd.

Dorucu, M., Crompton, D. W. T., Huntingford, F. A. & Walters, D. E. (1995). The ecology of endoparasitic helminth infections of brown trout (*Salmo trutta*) and rainbow trout (*Oncorhynchus mykiss*) in Scotland. *Folia Parasitologica*, **42**, 29–35.

Dove, A. D. M. (2000). Richness patterns in the parasite communities of exotic poeciliid fishes. *Parasitology*, **120**, 609–23.

Dudinak, V. & Snabel, V. (2001). Comparative analysis of Slovak and Czech population of *Pomphorhynchus laevis* (Acanthocephala) using morphological and isoenzyme analyses. *Acta Zoologica Universitatis Comenianae*, **44**, 41–50.

Duerr, H. P., Dietz, K. & Eichner, M. (2003). On the interpretation of age-intensity profiles and dispersion patterns in parasitological surveys. *Parasitology*, **126**, 87–101.

Dunagan, T. T. & Miller, D. M. (1980). *Macracanthorhynchus hirudinaceus* from swine: an eighteen-year record of Acanthocephala from Southern Illinois. *Proceedings of the Helminthological Society of Washington*, **47**, 33–6.

Edmonds, S. J. (1966). Hatching of eggs of *Moniliformis dubius*. *Experimental Parasitology*, **19**, 216–26.

Edwards, D. D. & Bush, A. O. (1989). Helminth communities in avocets: importance of the compound community. *Journal of Parasitology*, **75**, 225–38.

El-Darsh, H. E. M. & Whitfield, P. J. (1999). The parasite community infecting flounders, *Platichthys flesus*, in the tidal Thames. *Journal of Helminthology*, **73**, 203–14.

Elkins, A. A. & Nickol, B. B. (1983). The epizootiology of *Macracanthorhynchus ingens* in Louisiana. *Journal of Parasitology*, **69**, 951–6.

Esch, G. W., Hazen, T. C. & Aho, J. M. (1977). Parasitism and r- and K-selection. In *Regulation of Parasite Populations*, ed. G. W. Esch. New York: Academic Press, pp. 9–63.

Esch, G. W., Kennedy, C. R., Bush, A. O. & Aho, J. M. (1988). Patterns in helminth communities in freshwater fish in Great Britain: alternative strategies for colonisation. *Parasitology*, **96**, 519–32.

Esch, G. W., Bush, A. O. & Aho, J. M. (eds.) (1990). *Parasite Communities: Patterns and Processes*. London: Chapman & Hall.

Evans, D. W., Mathews, M. A. & McClintock, C. A. (2001). First record of *Pomphorhynchus laevis* (Acanthocephala) in fishes from Northern Ireland. *Journal of Fish Biology*, **59**, 166–8.

Ewald, J. A. & Nickol, B. B. (1989). Availability of caecal habitat as a density-dependent limit on survivorship of *Leptorhynchoides thecatus* in green sunfish, *Lepomis cyanellus*. *Parasitology*, **98**, 447–50.

Ewald, J. A., Crompton, D. W. T., Johnson, I. & Stoddart, R. C. (1991). The occurrence of *Centrorhynchus* (Acanthocephala) in shrews (*Sorex araneus* and *Sorex minutus*) in the United Kingdom. *Journal of Parasitology*, **77**, 485–7.

Ezenwa, V. O. (2003). Habitat overlap and gastrointestinal parasitism in sympatric African bovids. *Parasitology*, **126**, 379–88.

Farzaliev, A. M. & Petrochenko, V. I. (1980). New data on the life cycles of the acanthocephalan *Macracanthorhynchus catulinus* Kostylew, 1927 (Acanthocephala), a parasite of carnivores. *Trudy Vsesoyuznogo Instituta Gel'mintologii im K.I. Skryabina*, **25**, 140–4.

Fellis, K. J., Negovatich, N. J., Esch, G. W., Horak, I. G. & Boomber, J. (2003). Patterns of association, nestedness and species co-occurrence of helminth parasites in the greater kudu, *Tragelaphus strepsicorus*, in the Kruger National Park, South Africa, and the Etosha National Park, Namibia. *Journal of Parasitology*, **89**, 899–907.

Fitzgerald, R. D. & Mulcahy, M. F. (1983). Parasites of salmon *Salmo salar* L. and trout *Salmo trutta* L. in the River Shournagh. *Irish Fisheries Investigations Series A*, **23**, 24–31.

Font, W. F. (1997). Improbable colonists: helminth parasites of freshwater fishes on an oceanic island. *Micronesica*, **30**, 105–15.

Font, W. F. (1998). Parasites in paradise: patterns of helminth distribution in Hawaiian stream fishes. *Journal of Helminthology*, **72**, 307–11.

Foronda, P., Del Castillo, A., Abreu, N., Figueruelo, E., Pinero, J. & Casanovoa, J. C. (2003). Parasitic helminths of the wild rabbit, *Oryctolagus cuniculus*, in different bioclimatic zones in Tenerife, Canary Islands. *Journal of Helminthology*, **77**, 305–9.

Foronda, P., Casanova, J. C., Figueruelo, E., Abreu, N. & Feliu, C. (2005). The helminth fauna of the barbary partridge *Alectoris barbara* in Tenerife, Canary Islands. *Journal of Helminthology*, **79**, 133–8.

Foster, G. W., Kinsella, J. M., Walters, E. L., Schrader, M. S. & Forrester, D. J. (2002). Parasitic helminths of red-bellied woodpeckers (*Melanerpes carolinus*) from the Apalachicola National Forest in Florida. *Journal of Parasitology*, **88**, 1140–2.

Fuentes, M. V., Saez, S., Trelis, M., Galan-Puchades, M. T. & Esteban, J. G. (2004). The helminth community of the wood mouse, *Apodemus sylvaticus*, in the Sierra Espuna, Murcia, Spain. *Journal of Helminthology*, **78**, 219–23.

Galli, P., Mariniello, L., Crosa, G., Ortis, M., Occhipinti Ambrogi, A. & D'Amelio, S. (1998). Populations of *Acanthocephalus anguillae* and *Pomphorhynchus laevis* in rivers with different pollution levels. *Journal of Helminthology*, **72**, 331–5.

Garey, R. J., Nonnemacher, M. R., Near, T. J. & Nadler, S. A. (1996). Molecular evidence for acanthocephalans as a sub-taxon of rotifers. *Journal of Molecular Evolution*, **43**, 45–54.

Gaston, K. (1994). *Rarity*. London: Chapman & Hall.

Gelnar, M., Koubkova, B., Plankova, H. & Jurajda, P. (1996). Report on metazoan parasites of fishes of the river Morava with remarks on the effects of water pollution. *Helminthologia*, **31**, 47–56.

Gibson, D. I. (1972). Flounder parasites as biological tags. *Journal of Fish Biology*, **4**, 1–9.

Gleason, L. N. (1987). Population dynamics of *Pomphorhynchus bulbocolli* in *Gammarus pseudolimnaeus*. *Journal of Parasitology*, **73**, 1099–101.

Goater, C. P. (1989). Patterns of helminth parasitism in the oystercatcher, *Haematopus ostralegus*, from the Exe estuary, England. Unpublished Ph.D. thesis: University of Exeter.

Goater, C. P. & Bush, A. O. (1988). Intestinal helminth communities in long-billed curlew: the importance of congeneric host specialists. *Holarctic Ecology*, **11**, 140–5.

Golvan, Y. J. (1957). La specificité parasitaire chez les Acanthocephales. In *First Symposium on Host Specificity Among Parasites of Vertebrates*. Neuchatel: Paul Attinger, pp. 244–54.

Grey, A. J. & Hayunga, E. G. (1980). Evidence for alternative site selection by *Glaridacris laruei* (Cestoidea: Caryophyllidea) as a result of interspecific competition. *Journal of Parasitology*, **66**, 371–2.

Griffith, J. E., Beveridge, I., Chilton, N. B. & Johnson, P. M. (2000). Helminth communities of pademelons, *Thylogale stigmatica* and *T. thetis* from eastern Australia and Papua New Guinea. *Journal of Helminthology*, **74**, 307–14.

Guillen-Hernandez, S. & Whitfield, P. J. (2001). A comparison of freshwater and marine strains of *Pomphorhynchus laevis* occurring sympatrically in flounder, *Platichthys flesus*, in the tidal Thames. *Journal of Helminthology*, **75**, 237–43.

Guillen-Hernandez, S. & Whitfield, P. J. (2004). Intestinal helminth parasites in flounder *Platichthys flesus* from the River Thames: an infracommunity analysis. *Journal of Helminthology*, **78**, 297–303.

Hanski, I. & Gilpin, M. E. (1991). Metapopulation dynamics: brief history and conceptual domain. *Biological Journal of the Linnean Society*, **42**, 3–16.

Harms, C. E. (1965). The life cycle and larval development of *Octospinifer macilentis* (Acanthocephala: Neoechinorhynchidae). *Journal of Parasitology*, **51**, 286–93.

Harris, J. E. (1972). The immune response of a cyprinid fish to infections of the acanthocephalan *Pomphorhynchus laevis*. *International Journal for Parasitology*, **2**, 459–69.

Hart, B. L. (1994). Behavioural defence against parasites: interaction with parasite invasiveness. *Parasitology*, **109**, S139–S151.

Hartvigsen, R. & Kennedy, C. R. (1993). Patterns in the composition and richness of helminth communities in brown trout *Salmo trutta* in a group of reservoirs. *Journal of Fish Biology*, **43**, 603–15.

Haye, P. A. & Ojeda, F. P. (1998). Metabolic and behavioural alterations in the crab *Hemigrapsus crenulatus* (Milne-Edwards 1837) induced by its acanthocephalan parasite *Profilicollis antarcticus* (Zdzitowiecki 1985). *Journal of Experimental Marine Biology and Ecology*, **228**, 73–82.

Helle, E. & Valtonen, E. T. (1980). On the occurrence of *Corynosoma* spp. (Acanthocephala) in ringed seals *Pusa hispida* in the Bothnian Bay, Finland. *Canadian Journal of Zoology*, **58**, 298–303.

Helle, E. & Valtonen, E. T. (1981). Comparison between spring and autumn infection by *Corynosoma* (Acanthocephala) in the ringed seal *Pusa hispida* in the Bothnian Bay of the Baltic Sea. *Parasitology*, **82**, 287–96.

Hewitt, G. C. & Hine, P. M. (1972). Checklist of parasites of New Zealand fishes and of their hosts. *New Zealand Journal of Marine and Freshwater Research*, **6**, 69–114.

Hindsbo, O. (1972). Effects of *Polymorphus* (Acanthocephala) on colour and behaviour of *Gammarus lacustris*. *Nature, London*, **238**, 333.

Hine, P. M. (1978). Distribution of some parasites of freshwater eels in New Zealand. *New Zealand Journal of Marine and Freshwater Research*, **12**, 179–87.

Hine, P. M. & Kennedy, C. R. (1974a). Observations on the distribution, specificity and pathogenicity of the acanthocephalan *Pomphorhynchus laevis* (Muller). *Journal of Fish Biology*, **6**, 521–35.

Hine, P. M. & Kennedy, C. R. (1974b). The population biology of the acanthocephalan *Pomphorhynchus laevis* (Muller) in the River Avon. *Journal of Fish Biology*, **6**, 665−79.

Hnath, J. G. (1969). Transfer of an adult acanthocephalan from one fish host to another. *Transactions of the American Fisheries Society*, **98**, 332.

Hoffman, G. L. (1999). *Parasites of North American Freshwater Fishes*, 2nd edn. New York: Cornell University Press.

Holland, C. (1984). Interactions between *Moniliformis* (Acanthocephala) and *Nippostrongylus* (Nematoda) in the small intestine of laboratory rats. *Parasitology*, **88**, 303−15.

Holland, C. (1987). Interspecific effects between *Moniliformis* (Acanthocephala), *Hymenolepis* (Cestoda) and *Nippostrongylus* (Nematoda) in the laboratory rat. *Parasitology*, **94**, 567−81.

Holland, C. V. (1983). Interactions between three species of helminth parasite in the rat small intestine. Unpublished Ph.D. thesis: University of Cambridge.

Holland, C. V. & Kennedy, C. R. (1997). A checklist of parasitic helminth and crustacean species recorded in freshwater fish from Ireland. *Biology and Environment: Proceedings of the Royal Irish Academy*, **97B**, 225−43.

Holmes, J. C. (1961). Effects of concurrent infections on *Hymenolepis diminuta* (Cestoda) and *Moniliformis dubius* (Acanthocephala): I. General effects and comparison with crowding. *Journal of Parasitology*, **47**, 209−16.

Holmes, J. C. (1962a). Effects of concurrent infections on *Hymenolepis diminuta* (Cestoda) and *Moniliformis dubius* (Acanthocephala): II. Growth. *Journal of Parasitology*, **48**, 87−96.

Holmes, J. C. (1962b). Effects of concurrent infections on *Hymenolepis diminuta* (Cestoda) and *Moniliformis dubius* (Acanthocephala): III. Effects in hamsters. *Journal of Parasitology*, **48**, 97−100.

Holmes, J. C. (1973). Site selection by parasitic helminths: interspecific interactions, site segregation, and their importance to the development of helminth communities. *Canadian Journal of Zoology*, **51**, 333−47.

Holmes, J. C. (1982). Impact of infectious disease agents on the population growth and geographical distribution of animals. In *Population Biology of Infectious Diseases*, ed. R. M. Anderson & R. M. May, Berlin: Springer Verlag, pp. 37−51.

Holmes, J. C. (1987). Helminth communities of parasites. *International Journal for Parasitology*, **17**, 203−8.

Holmes, J. C. (1990). Helminth communities in marine fishes. In *Parasite Communities: Patterns and Processes*, ed. G. W. Esch, A. O. Bush & J. M. Aho. London & New York: Chapman & Hall, pp. 101−30.

Holmes, J. C. & Bartoli, P. (1993). Spatio-temporal structure of the communities of helminths in the digestive tract of *Sciaena umbra* L. 1758. *Parasitology*, **106**, 519−26.

Holmes, J. C. & Price, P. W. (1986). Communities of parasites. In *Community Ecology: Pattern and Process* ed. J. Kikkawa & D. J. Anderson. Oxford: Blackwell Scientific Publications, pp. 187−213.

Holmes, J. C., Hobbs, R. P. & Leong, T. S. (1977). Populations in perspective: community organisation and regulation of parasites populations. In *Regulation of Parasite Populations*, ed. G. W. Esch. New York: Academic Press, pp. 209−45.

Hopp, W. B. (1954). Studies on the morphology and life cycle of *Neoechinorhynchus emydis* (Leidy), an acanthocephalan parasite of the map turtle, *Graptemys geographica* (LeSueur). *Journal of Parasitology*, **40**, 284−99.

Hudson, P. & Greenman, J. (1998). Competition mediated by parasites: biological and theoretical progress. *Trends in Ecology and Evolution*, **13**, 387–90.

Huffman, D. G. & Bullock, W. L. (1975). Meristograms: graphical analysis of serial variations of proboscis hooks of *Echinorhynchus* (Acanthocephala). *Systematic Zoology*, **24**, 333–45.

Hurlbert, I. A. R. & Boag, B. (2001). The potential role of habitat on intestinal helminths of mountain hares, *Lepus timidus*. *Journal of Helminthology*, **75**, 345–9.

Hynes, H. B. N. & Nicholas, W. L. (1958). The resistance of *Gammarus* spp. to infection by *Polymorphus minutus* (Goeze, 1782) (Acanthocephala). *Annals of Tropical Medicine and Parasitology*, **52**, 376–83.

Hynes, H. B. N. & Nicholas, W. L. (1963). The importance of the acanthocephalan *Polymorphus minutus* as a parasite of domestic ducks in the United Kingdom. *Journal of Helminthology*, **37**, 185–98.

Ibragimov, Sh. R. (1985). Parasite fauna of acipenserid fish of the Caspian Sea. *Izvestia Academii Nauk Azerbaidzhan*, **2**, 47–51.

Itamies, J., Valtonen, E. T. & Fagerholm, H. P. (1980). *Polymorphus minutus* infection in eiders and its role as a possible cause of death. *Annales Zoologici Fennici*, **17**, 285–9.

Jackson, J. A. & Nickol, B. B. (1981). Ecology of *Mediorhynchus centurorum* host specificity. *Journal of Parasitology*, **65**, 167–9.

Janovy, J., Jr. & Hardin, E. L. (1988). Diversity of the parasite assemblage of *Fundulus zebrinus* in the Platte River of Nebraska. *Journal of Parasitology*, **74**, 207–13.

Jensen, T. (1952). The life cycle of the fish acanthocephalan, *Pomphorhynchus bulbocolli* (Linkins) Van Cleave, 1919, with some observations on larval development in vitro. Unpublished Ph.D. thesis, University of Minnesota.

Jones, A., Bailey, T. A., Nothelfer, H. B., *et al.* (1996). Parasites of wild houbara bustards in the United Arab Emirates. *Journal of Helminthology*, **70**, 21–5.

Karvonen, A., Bagge, A. M. & Valtonen, E. T. (2005). Parasite assemblages of crucian carp (*Carassius carassius*): is depauperate composition explained by lack of parasite exchange, extreme environmental conditions or host unsuitability? *Parasitology*, **131**, 273–8.

Kates, K. C. (1942). Viability of the eggs of the swine thorn-headed worm (*Macracanthorhynchus hirudinaceus*). *Journal of Agricultural Research*, **64**, 93–100.

Kates, K. C. (1944). Some observations on experimental infections of pigs with the thorn-headed worm *Macracanthocephalus hirudinaceus*. *American Journal of Veterinary Research*, **5**, 166–72.

Kennedy, C. R. (1972). The effects of temperature and other factors upon the establishment and survival of *Pomphorhynchus laevis* (Acanthocephala) in goldfish (*Carassius auratus*). *Parasitology*, **65**, 283–94.

Kennedy, C. R. (1974). The importance of parasite mortality in regulating the population size of the acanthocephalan *Pomphorhynchus laevis* in goldfish. *Parasitology*, **68**, 271–84.

Kennedy, C. R. (1975). *Ecological Animal Parasitology*. Oxford: Blackwell Scientific Publications.

Kennedy, C. R. (1976). Reproduction and dispersal. In *Ecological Aspects of Parasitology*, ed. C. R. Kennedy. Amsterdam: North Holland, pp. 143–60.

Kennedy, C. R. (1977). Distributional and zoogeographical characteristics of the parasite fauna of the char *Salvelinus alpinus* in Arctic Norway, including Spitsbergen and Jan Mayen Islands. *Astarte*, **10**, 49–55.

Kennedy, C. R. (1978a). The parasite fauna of resident char *Salvelinus alpinus* from Arctic Islands, with special reference to Bear Island. *Journal of Fish Biology,* **13,** 457–66.

Kennedy, C. R. (1978b). An analysis of the metazoan parasitocoenosis of brown trout *Salmo trutta* from British lakes. *Journal of Fish Biology,* **13,** 255–63.

Kennedy, C. R. (1984a). The status of flounders, *Platichthys flesus* (L.), as hosts of the acanthocephalan *Pomphorhynchus laevis* (Muller) and its survival in marine conditions. *Journal of Fish Biology,* **24,** 135–49.

Kennedy, C. R. (1984b). Dynamics of a declining population of the acanthocephalan *Acanthocephalus clavula* in eels *Anguilla anguilla* in a small river. *Journal of Fish Biology,* **25,** 665–77.

Kennedy, C. R. (1985a). Regulation and dynamics of acanthocephalan populations. In *Biology of the Acanthocephala,* ed. D. W. T. Crompton & B. B. Nickol, Cambridge: Cambridge University Press, pp. 385–416.

Kennedy, C. R. (1985b). Site segregation by species of Acanthocephala in fish with special reference to eels. *Parasitology,* **90,** 375–90.

Kennedy, C. R. (1990). Helminth communities in freshwater fish: structured communities or stochastic assemblages? In *Parasite Communities: Patterns and Processes,* ed. G. W. Esch, A. O. Bush & J. M. Aho, London & New York: Chapman & Hall, pp. 130–56.

Kennedy, C. R. (1992). Field evidence for interactions between the acanthocephalans *Acanthocephalus anguillae* and *A. lucii* in eels, *Anguilla anguilla.* *Ecological Parasitology,* **1,** 122–34.

Kennedy, C. R. (1993a). Acanthocephala. In *Asexual Reproduction and Reproductive Biology of Invertebrates,* ed. A. D. Adiyodi & K. G. Adiyodi, Vol. 6A. New Delhi: Oxford and IBH, pp. 279–95.

Kennedy, C. R. (1993b). The dynamics of helminth communities in eels *Anguilla anguilla* in a small stream: long term changes in richness and structure. *Parasitology,* **107,** 71–8.

Kennedy, C. R. (1993c). Introductions, spread and colonization of new localities by fish helminth and crustacean parasites in the British Isles: a perspective and appraisal. *Journal of Fish Biology,* **43,** 287–301.

Kennedy, C. R. (1994). The ecology of introductions. In *Parasitic Diseases of Fish,* ed. A. W. Pike & J. W. Lewis. Tresaith Cardigan: Samara, pp. 198–208.

Kennedy, C. R. (1995). Richness and diversity of macroparasite communities in tropical eels *Anguilla reinhardtii* in Queensland, Australia. *Parasitology,* **111,** 233–45.

Kennedy, C. R. (1996). Colonisation and establishment of *Pomphorhynchus laevis* (Acanthocephala) in an isolated English river. *Journal of Helminthology,* **70,** 27–31.

Kennedy, C. R. (1997a). Long-term and seasonal changes in composition and richness of intestinal helminth communities in eels *Anguilla anguilla* of an isolated English river. *Folia Parasitolgica,* **44,** 267–73.

Kennedy, C. R. (1997b). Freshwater fish parasites and environmental quality: an overview and caution. *Parassitologia,* **39,** 249–54.

Kennedy, C. R. (1998). Aquatic birds as agents of parasite dispersal: a field test of the effectiveness of helminth colonisation strategies. *Bulletin of the Scandinavian Society for Parasitology,* **8,** 23–8.

Kennedy, C. R. (1999). Post-cyclic transmission in *Pomphorhynchus laevis* (Acanthocephala). *Folia Parasitologica,* **46,** 111–16.

Kennedy, C. R. (2000). Freshwater parasites and habitat change. *Bulletin of the Scandinavian Society for Parasitology,* **10,** 39–43.

Kennedy, C. R. (2001). Metapopulation and community dynamics of helminth parasites of eels *Anguilla anguilla* in the River Exe system. *Parasitology*, **122**, 689–98.

Kennedy, C. R. (2003). Evolution of host-parasite systems in the Acanthocephala: speciation and scale in the genus *Pomphorhynchus*. In *Taxonomie, Ecologie et Evolution des Metazoaires Parasites*, ed. C. Combes and J. Jourdane, Tome II. Perpignan: Presses Universitaires de Perpignan, pp. 11–35.

Kennedy, C. R. & Burrough, R. J. (1978). Parasites of trout and perch in Malham Tarn. *Field Studies*, **4**, 617–29.

Kennedy, C. R. & Bush, A. O. (1992). Species richness in helminth communities: the importance of multiple congeners. *Parasitology*, **104**, 189–97.

Kennedy, C. R. & Bush, A. O. (1994). The relationship between pattern and scale in parasite communities: a stranger in a strange land. *Parasitology*, **109**, 187–96.

Kennedy, C. R. & Guégan, J. F. (1994). Regional versus local parasite richness in British freshwater fish: saturated or unsaturated communities? *Parasitology*, **109**, 175–85.

Kennedy, C. R. & Guégan, J. F. (1996). The number of niches in intestinal helminth communities of *Anguilla anguilla*: are there enough spaces for parasites? *Parasitology*, **113**, 293–302.

Kennedy, C. R. & Hartvigsen, R. A. (2000). Richness and diversity of intestinal metazoan communities in brown trout *Salmo trutta* compared to those of eels *Anguilla anguilla* in their European heartlands. *Parasitology*, **121**, 55–64.

Kennedy, C. R. & Lord, D. (1982). Habitat specificity of the acanthocephalan *Acanthocephalus clavula* (Dujardin, 1845) in eels *Anguilla anguilla* (L.). *Journal of Helminthology*, **56**, 121–9.

Kennedy, C. R. & Moriarty, C. (1987). Co-existence of congeneric species of Acanthocephala: *Acanthocephalus lucii* and *A. anguillae* in eels *Anguilla anguilla* in Ireland. *Parasitology*, **95**, 301–10.

Kennedy, C. R. & Moriarty, C. (2002). Long-term stability in the richness and structure of helminth communities in eels *Anguilla anguilla* in Lough Derg, River Shannon, Ireland. *Journal of Helminthology*, **76**, 315–22.

Kennedy, C. R. & Pojmanska, T. (1996). Richness and diversity of helminth communities in common carp and in three more recently introduced carp species. *Journal of Fish Biology*, **48**, 89–100.

Kennedy, C. R. & Rumpus, A. (1977). Long term changes in the size of the *Pomphorhynchus laevis* (Acanthocephala) population in the River Avon. *Journal of Fish Biology*, **10**, 35–42.

Kennedy, C. R., Broughton, P. F. & Hine, P. M. (1976). The sites occupied by the acanthocephalan *Pomphorhynchus laevis* in the alimentary canal of fish. *Parasitology*, **72**, 195–206.

Kennedy,C. R., Broughton, P. F. & Hine, P. M. (1978). The status of brown and rainbow trout, *Salmo trutta* and *S. gairdneri* as hosts of the acanthocephalan *Pomphorhynchus laevis*. *Journal of Fish Biology*, **13**, 265–75.

Kennedy, C. R., Bush, A. O. & Aho, J. M. (1986a). Patterns in helminth communities: why are birds and fish different? *Parasitology*, **93**, 205–15.

Kennedy, C. R., Laffoley, D. d'A., Bishop, G., Taylor, M. & Jones, P. (1986b). Structure and organisation of parasite communities of freshwater fish of Jersey, Channel Islands. *Journal of Fish Biology*, **29**, 215–26.

Kennedy, C. R., Bates, R. M. & Brown, A. F. (1989). Discontinuous distributions of the fish acanthocephalans *Pomphorhynchus laevis* and *Acanthocephalus anguillae* in Britain and Ireland: a hypothesis. *Journal of Fish Biology*, **34**, 607–19.

Kennedy, C. R., Hartvigsen, R. & Halvorsen, O. (1991). The importance of fish stocking in the dissemination of parasites throughout a group of reservoirs. *Journal of Fish Biology*, **38**, 541–52.

Kennedy, C. R., Nie, P., Kaspers, J. & Paulisse, J. (1992). Are eels (*Anguilla anguilla* L.) planktonic feeders? Evidence from parasite communities. *Journal of Fish Biology*, **41**, 567–80.

Kennedy, C. R., Di Cave, D., Berilli, F. & Orecchia, P. (1997). Composition and structure of helminth communities in eels *Anguilla anguilla* from Italian coastal lagoons. *Journal of Helminthology*, **71**, 35–40.

Kennedy, C. R., Berrili, F., Di Cave, D., De Liberato, C. & Orecchia, P. (1998). Composition and diversity of helminth communities in eels *Anguilla anguilla* in the River Tiber: long-term changes and comparison with insular Europe. *Journal of Helminthology*, **72**, 301–6.

Keymer, A. E. & Read, A. F. (1991). Behavioural ecology: the impact of parasitism. In *Parasite–Host Associations: Coexistence or Conflict?*, ed. C. A. Toft, A. Aeschlimann and L. Bolis. Oxford: Oxford University Press, pp. 37–61.

Khan, R. A. & Payne, J. F. (1997). A multidisciplinary approach using several biomarkers, including a parasite, as indicators of pollution: a case history from a paper mill in Newfoundland. *Parassitologia*, **39**, 183–8.

Komarova, M. S. (1950). On the problem of the life cycle of *Acanthocephalus lucii*. *Doklady Akademi Nauka SSSR*, **70**, 359–60.

Kral'ova-Hromadova, I., Tietz, D. F., Shinn, A. P. & Spakulova, M. (2003). ITS rDNA sequences of *Pomphorhynchus laevis* (Zoega in Muller, 1776) and *P. lucyi* Williams and Rogers, 1984 (Acanthocephala: Palaeacanthocephala). *Systematic Parasitology*, **56**, 141–5.

Kulachkova, V. G. & Timofeeva, T. A. (1977). The acanthocephalan *Echinorhynchus gadi* (Zoega) from relict cod in Lake Mogil'noe. *Parazitologiya*, **11**, 316–20.

Kussat, R. H. (1969). A comparison of aquatic communities in the Bow river above and below sources of domestic and industrial wastes from the city of Calgary. *Canadian Fish Culture*, **40**, 3–31.

Lackie, A. M. & Holt, H. F. (1988). Immunosuppression by larvae of *Moniliformis moniliformis* (Acanthocephala) in their cockroach host *Periplaneta americana*. *Parasitology*, **98**, 307–14.

Lackie, J. M. (1972). The course of infection and growth of *Moniliformis dubius* (Acanthocephala) in the intermediate host *Periplaneta americana*. *Parasitology*, **64**, 95–106.

Lackie, J. M. (1975). The host specificity of *Moniliformis dubius* (Acanthocephala) a parasite of cockroaches. *International Journal for Parasitology*, **5**, 301–7.

Lafferty, K. D. (1992). Foraging on prey that are modified by parasites. *American Naturalist*, **140**, 854–67.

Laimgruber, S., Schludermann, C., Konecny, R. & Chovanec, A. (2005). Helminth communities of the barbel *Barbus barbus* from large river systems in Austria. *Journal of Helminthology*, **79**, 143–9.

Lassiere, O. L. (1988). Host-parasite relationships between larval *Sialis lutaria* (Megaloptera) and *Neoechinorhynchus rutili* (Acanthocephala). *Parasitology*, **97**, 331–8.

Lassiere, O. L. & Crompton, D. W. T. (1988). Evidence for post-cyclic transmission in the life-history of *Neoechinorhynchus rutili* (Acanthocephala). *Parasitology*, **97**, 339–43.

Latham, A. D. M. & Poulin, R. (2001). Effect of acanthocephalan parasites on the behaviour and colouration of the mud crab *Macrophthalmus hirtipes* (Brachyura: Ocypodidae). *Marine Biology*, **139**, 1147–54.

Latham, A. D. M. & Poulin, R. (2002). Effect of acanthocephalan parasites on hiding behaviour in two species of shore crabs. *Journal of Helminthology*, **76**, 323–6.

Leadabrand, C. C. & Nickol, B. B. (1993). Establishment, survival, site selection and development of *Leptorhynchoides thecatus* in largemouth bass, *Micropterus salmoides*. *Parasitology*, **106**, 495–501.

Leong, T. S. & Holmes, J. C. (1981). Communities of metazoan parasites in open water fishes of Cold Lake, Alberta. *Journal of Fish Biology*, **18**, 693–713.

Liat, L. B. & Pike, A. W. (1980). The incidence and distribution of *Profilicollis botulus* (Acanthocephala) in the eider duck, *Somateria molissima*, and in its intermediate host the shore crab, *Carcinus maenus*, in north east Scotland. *Journal of Zoology, London*, **190**, 39–51.

Loker, E. S. (1994). On being a parasite in an invertebrate host: a short survival course. *Journal of Parasitology*, **80**, 728–47.

Lotz, J. M. & Font, W. F. (1994). Excess positive associations in communities of intestinal helminths of bats: a refined null hypothesis and a test of the facilitation hypothesis. *Journal of Parasitology*, **80**, 398–413.

Lyndon, A. R. (1996). The role of acanthocephalan parasites in the predation of freshwater isopods by fish. In *Aquatic Predators and their Prey*, ed. P. R. Greenstreet & M. L. Tasker. Oxford: Fishing News Books, Blackwell Scientific, pp. 26–32.

Lyndon, A. R. & Kennedy, C. R. (2001). Colonisation and extinction in relation to competition and resource partitioning in acanthocephalans of freshwater fishes of the British Isles. *Folia Parasitologica*, **48**, 37–46.

MacArthur, R. H. (1972). *Geographical Ecology: Patterns in the Distribution of Species*. New York: Harper & Row.

MacArthur, R. H. & Wilson, E. O. (1967). *The Theory of Island Biogeography*. Princeton: Princeton University Press.

MacKenzie, K. (1983). Parasites as biological tags in fish population studies. *Advances in Applied Biology*, **7**, 251–331.

MacKenzie, K. (1986). Parasites as indicators of host populations. In *Parasitology– Quo vadit? Proceedings of the Sixth International Congress of Parasitology* ed. M. J. Howell. Canberra: Australian Academy of Science, pp. 345–52.

MacKenzie, K. (2002). Parasites as biological tags in population studies of marine organisms: an update. *Parasitology*, **124**, S153–S163.

MacKenzie, K., Williams, H. H., Williams, B., McVicar, A. H. & Siddal, R. (1995). Parasites as indicators of water quality and the potential use of helminth transmission in marine pollution studies. *Advances in Parasitology*, **35**, 85–144.

Madhavi, R. & Sai Ram, B. K. (2000). Community structure of helminth parasites of the tuna, *Euthymnnus affinis*, from the Visakhapatnam coast, Bay of Bengal. *Journal of Helminthology*, **74**, 337–42.

Mafiana, C. F., Osho, M. B. & Sam-Wobo, S. (1997). Gastrointestinal helminth parasites of the black rat (*Rattus rattus*) in Abeokuta, southwest Nigeria. *Journal of Helminthology*, **71**, 217–20.

Marcogliese, D. J. (2002). Food webs and the transmission of parasites to marine fish. *Parasitology*, **124**, S83–S99.

Marcogliese, D. J. & Cone, D. K. (1993). What metazoan parasites tell us about the evolution of American and European eels. *Evolution*, **47**, 1632–5.

Marcogliese, D. J. & Cone, D. K. (1996). On the distribution and abundance of eel parasites in Nova Scotia: influence of pH. *Journal of Parasitology*, **82**, 389–99.

Marcogliese, D. J. & Cone, D. K. (1997a). Food webs: a plea for parasites. *Trends in Ecology and Evolution*, **12**, 320–5.

Marcogliese, D. J. & Cone, D. K. (1997b). Parasite communities as indicators of ecosystem stress. *Parassitologia*, **39**, 227–32.

Marcogliese, D. J. & Cone, D.K (1998). Comparison of richness and diversity of macroparasite communities among eels from Nova Scotia, the United Kingdom and Australia. *Parasitology*, **116**, 73–83.

Margolis, L. (1963). Parasites as indicators of the geographcal origin of sockeye salmon, *Oncorhynchus nerka* (Walbaum), occurring in the North Pacific Ocean and adjacent areas. INPFC Document No. 466. *Fisheries Research Board of Canada, Bulletin*, **11**, 101–56.

Margolis, L. (1965). Parasites as an auxilliary source of information about the biology of Pacific salmon (Genus *Oncorhynchus*). *Journal of the Fisheries Research Board of Canada*, **22**, 1387–95.

Martin, J. E. & Roca, V. (2004). Helminth infracommunities of a population of the Gran Canaria giant lizard *Gallotia stehlini*. *Journal of Helminthology*, **78**, 319–22.

Mason, C. F. (1991). *Biology of Freshwater Pollution*, 2nd edn. Harlow: Longman Scientific & Technical.

Matsuo, K. & Oku, Y. (2002). Endoparasites of three species of house geckoes in Lampung, Indonesia. *Journal of Helminthology*, **76**, 53–7.

May, R. M. & Anderson, R. M. (1978). Regulation and stability of host-parasite population interactions: II. Destabilising processes. *Journal of Animal Ecology*, **47**, 249–68.

Maynard, B. J., Wellnitz, T. A., Zanini, N., Wright, W. G. & Dezfuli, B. S. (1998). Parasite-altered behaviour in a crustacean intermediate host: field and laboratory studies. *Journal of Parasitology*, **84**, 1102–6.

Mazzi, D. & Bakker, T. C. M. (2003). A predator's dilemma: prey choice and parasite susceptibility in three-spined sticklebacks. *Parasitology*, **126**, 339–47.

McCahon, C. P., Maund, S. J. & Poulton, M. J. (1991). The effect of the acanthocephalan parasite *Pomphorhynchus laevis* on the drift of its intermediate host *Gammarus pulex*. *Freshwater Biology*, **25**, 507–13.

McCormick, A. L. & Nickol, B. B. (2004). Postcyclic transmission and its effect on the distribution of *Paulisentis missouriensis* (Acanthocephala) in the definitive host *Semotilus atromaculatus*. *Journal of Parasitology*, **90**, 103–7.

McDonald, T. E & Margolis, L. (1995). Synopsis of the parasites of fishes of Canada: Supplement (1978–1993). *Canadian Special Publications of Fisheries and Aquatic Sciences*, **122**, 1–265.

McVicar, A. H. (1997). The development of marine environmental monitoring using fish diseases. *Parassitologia*, **39**, 177–81.

Menezes, V. A., Vrcibradic, D., Vicente, J. J., Dutra, G. F. & Rocha, C. F. D. (2004). Helminths infecting the parthenogenetic whiptail lizard *Cnemidophorus nativo* in a restinga habitat on Bahia State, Brazil. *Journal of Helminthology*, **78**, 323–8.

Merritt, S. V. & Pratt, I. (1964). The life history of *Neoechinorhynchus rutili* and its development in the intermediate host (Acanthocephala: Neoechinorhynchidae). *Journal of Parasitology*, **50**, 394–400.

Meyer, A. (1938). Klasse: Acanthocephala, Acanthozephalen, Kratzer. In *Die Tierwelt Mitteleuropas*, ed. P. Bromher, P. Ehrmann & G. Ulmer, Band 1, Lief. 6, Leipzig: Verlag Von Quelle & Meyer, pp. 1–40.

Molloy, S., Holland, C. & Poole, R. (1993). Helminth parasites of brown and sea trout *Salmo trutta* L. from the west coast of Ireland. *Biology and Environment: Proceedings of the Royal Irish Academy* **93B**, 137–42.

Molloy, S., Holland, C. & O'Regan, M. (1995). Population biology of *Pomphorhynchus laevis* in brown trout from two lakes in the west of Ireland. *Journal of Helminthology*, **69**, 229–35.

Molnar, K. & Szekely, C. (1995). Parasitological survey of some important fish species of Lake Balaton. *Parasitologia Hungaria*, **28**, 63–82.

Moore, D. V. (1946). Studies on the life history and development of *Moniliformis dubius* Meyer, 1933. *Journal of Parasitology*, **32**, 257–71.

Moore, J. K. (1983a). Responses of an avian predator and its isopod prey to an acanthocephalan parasite. *Ecology*, **64**, 1000–15.

Moore, J. K. (1983b). Altered behaviour in cockroaches (*Periplaneta americana*) infected with an acanthocephala, *Moniliformis moniliformis*. *Journal of Parasitolgy*, **69**, 1174–6.

Moore, J. K. (1984). Altered behavioural responses in intermediate hosts: an acanthocephalan parasite strategy. *American Naturalist*, **123**, 572–7.

Moore, J. K. & Bell, D. H. (1983). Pathology (?) of *Plagiorhynchus cylindraceus* in the starling, *Sternus vulgaris*. *Journal of Parasitology*, **69**, 387–90.

Moore, J. K. & Crompton, D. W. T. (1993). A quantitative study of the susceptibility of cockroach species to *Moniliformis moniliformis* (Acanthocephala). *Parasitology*, **107**, 63–9.

Moore, J. K. & Gotelli, N. J. (1990). Phylogenetic perspective on the evolution of altered host behaviours: a critical look at the manipulation hypothesis. In *Parasitism and Host Behaviour* ed. C. J. Barnard & J. M. Behnke, London: Taylor & Francis, pp. 193–229.

Moore, J. K. & Gotelli, N. J. (1996). Evolutionary patterns of altered behaviour and susceptibility in parasitized hosts. *Evolution*, **50**, 807–19.

Moore, J. K. & Simberloff, D. (1990). Gastrointestinal helminth communities of bob-white quail. *Ecology*, **71**, 344–59.

Moravec, F. (1985). Occurrence of endoparasitic helminths in eels (*Anguilla anguilla* L.) from the Macha Lake fishpond system, Czechoslovakia. *Folia Parasitologica*, **32**, 113–25.

Moravec, F. & Scholz, T. (1991). Observations on the biology of *Pomphorhynchus laevis* (Zoega in Muller, 1776) (Acanthocephala) in the Rokytna River, Czech and Slovak Federative Republic. *Helminthologia*, **28**, 23–9.

Moravec, F., Holcik, J. & Meszaros, J. (1989). Notes on the helminth fauna of the sterlet, *Acipenser ruthenus* L., in Slovakia. *Biologia (Bratislava)*, **44**, 151–9.

Moravec, F., Konecny, R., Baska, F. *et al.* (1997). *Endohelminth Fauna of Barbel, Barbus barbus (L.), Under Ecological Conditions of the Danube Basin in Central Europe*. Prague: Academia.

Moro, P. L., Ballarta, J., Gilman, R. H., Jeguia, G., Rojas, M. & Montes, G. (1998). Intestinal parasites of the grey fox (*Pseudalopex culpaeus*) in the central Peruvian Andes. *Journal of Helminthology*, **72**, 87–9.

Moss, B. (1988). *Ecology of Fresh Waters*. 2nd edn. Oxford: Blackwell Scientific.

Mouritsen, K. N & Poulin, R. (2002). Parasitism, community structure and biodiversity in intertidal ecosystems. *Parasitology*, **124**, S101–S117.

Munro, M. A., Whitfield, P. J. & Diffley, R. (1989). *Pomphorhynchus laevis* (Muller) in the flounder, *Platichthys flesus* L., in the tidal River Thames: population structure, microhabitat utilization and reproductive status in the field and under conditions of controlled salinity. *Journal of Fish Biology*, **35**, 719–35.

Munro, M. A., Reid, A. & Whitfield, P. J. (1990). Genomic divergence in the ecologically differentiated English freshwater and marine strains of *Pomphorhynchus laevis* (Acanthocephala): Palaeacanthocephala: a preliminary investigation. *Parasitology*, **101**, 451–5.

Muzzall, P. M. (1980). Ecology and seasonal abundance of three acanthocephalan species infecting white suckers in SE New Hampshire. *Journal of Parasitology*, **66**, 127–33.

Muzzall, P. M. (1991). Helminth infra-communities of the frogs *Rana catesbeiana* and *Rana clamitans* from Turkey Marsh, Michigan. *Journal of Parasitology*, **77**, 366–71.

Muzzall, P. M. & Bullock, W. L. (1978). Seasonal occurrence and host-parasite relationships of *Neoechinorhynchus saginatus* Van Cleave & Bangham, 1949 in the fallfish *Semotilus corporalis* (Mitchell). *Journal of Parasitology*, **64**, 860–5.

Muzzall, P. M. & Rabalais, F. C. (1975). Studies on *Acanthocephalus jacksoni* Bullock 1962 (Acanthocephala: Echinorhynchidae). III. The altered behaviour of *Lirceus lineatus* (Say.) infected with cystacanths of *Acanthocephalus jacksoni*. *Proceedings of the Helminthological Society of Washington*, **42**, 116–18.

Nelson, M. J. & Nickol, B. B. (1986). Survival of *Macracanthorhynchus ingens* in swine and histopathology of infection in swine and racoons. *Journal of Parasitology*, **72**, 306–14.

Nicholas, W. L. & Hynes, H. B. N. (1958). Studies on *Polymorphus minutus* (Goeze, 1782) (Acanthocephala) as a parasite of the domestic duck. *Annals of Tropical Medicine and Parasitology*, **52**, 36–47.

Nickol, B. B. (1977). Life history and host specificity of *Mediorhynchus centrurorum* Nickol, 1969 (Acanthocephala: Gigantorhynchidae). *Journal of Parasitology*, **63**, 104–11.

Nickol, B. B. (1985). Epizootiology. In *Biology of the Acanthocephala*, ed. D. W. T. Crompton & B. B. Nickol. Cambridge: Cambridge University Press, pp. 307–46.

Nickol, B. B. (2003). Is postcyclic transmission underestimated as an epizootiological factor for acanthocephalans? *Helminthologia*, **40**, 93–5.

Nickol, B. B. & Dappen, G. E. (1982). *Armadillidium vulgare* (Isopoda) as an intermediate host of *Plagiorhynchus cylindraceus* (Acanthocephala) and isopod response to infection. *Journal of Parasitology*, **68**, 570–5.

Nickol, B. B. & Heard, R. W. (1973). Host parasite relationships of *Fessisentis necturorum* (Acanthocephala: Fessisentidae). *Proceedings of the Helminthological Society of Washington*, **40**, 204–8.

Nickol, B. B., Heard, R. W. & Smith, N. F. (2002a). Acanthocephalans from crabs in the southeastern U.S., with the first intermediate hosts known for *Arhythmorhynchus frassoni* and *Hexaglandula corynosoma*. *Journal of Parasitology*, **88**, 79–83.

Nickol, B. B., Helle, E. & Valtonen, E. T. (2002b). *Corynosoma magdaleni* in grey seals from the gulf of Bothnia, with amended descriptions of *Corynosoma strumosum* and *Corynosoma magdaleni*. *Journal of Parasitology*, **88**, 1222–9.

Nie, P. (1995). Communities of intestinal helminths of carp, *Cyprinus carpio*, in highland lakes in Yunnan province of southwest China. *Acta Parasitologica*, **40**, 148–51.

Norton, J., Lewis, J. W. & Rollinson, D. (2003). Parasite infracommunity diversity in eels: a reflection of local component community diversity. *Parasitology*, **127**, 475–82.

Norton, J., Lewis, J. W. & Rollinson, D. (2004a). Temporal and spatial patterns of nestedness in eel macroparasites communities. *Parasitology*, **129**, 203–11.

Norton, J., Rollinson, D. & Lewis, J. W. (2004b). Patterns of infracommunity species richness in eels *Anguilla anguilla*. *Journal of Helminthology*, **78**, 141–6.

Nwosu, C. O., Ogunrinade, A. F. & Fagbemi, B. O. (1996). Prevalence and seasonal changes in the gastrointestinal helminths of Nigerian goats. *Journal of Helminthology*, **70**, 329–33.

O'Connell, M. P. & Fires, J. M. (2004). Helminth communities of the lesser sand eel *Ammodytes tobianus* L., off the west coast of Ireland. *Journal of Parasitology*, **90**, 1058–61.

O'Mahony, E. M., Kennedy, C. R. & Holland, C. V. (2004a). Morphological characters distinguish Irish and English populations of the acanthocephalan *Pomphorhynchus laevis* (Muller, 1776). *Systematic Parasitology*, **59**, 147–57.

O'Mahony, E. M., Bradley, D. G., Kennedy, C. R. & Holland, C. V. (2004b). Evidence for the hypothesis of strain formation in *Pomphorhynchus laevis* (Acanthocephala): an investigation using DNA sequences. *Parasitology*, **129**, 341–7.

O'Neill, G. & Whelan, J. (2002). The occurrence of *Corynosoma strumosum* in the grey seal *Halichoerus grypus* caught off the Atlantic coast of Ireland. *Journal of Helminthology*, **76**, 231–4.

Oetinger, D. F. & Nickol, B. B. (1974). A possible function of the fibrillar coat in *Acanthocephalus jacksoni* eggs. *Journal of Parasitology*, **60**, 1055–6.

Oetinger, D. F. & Nickol, B. B. (1981). Effects of acanthocephalans on pigmentation of freshwater isopods. *Journal of Parasitology*, **67**, 672–84.

Oetinger, D. F. & Nickol, B. B. (1982). Developmental relationships between acanthocephalans and altered pigmentation in freshwater isopods. *Journal of Parasitology*, **68**, 463–9.

Olson, P. D. & Nickol, B. B. (1996). Comparison of *Leptorhynchoides thecatus* (Acanthocephala) recruitment into green sunfish and largemouth bass populations. *Journal of Parasitology*, **82**, 702–6.

Ortubay, S., Ubeda, C., Semenas, L. & Kennedy, C. R. (1991). *Pomphorhynchus patagonicus* n.sp. (Acanthocephala: Pomphorhynchidae) from freshwater fishes of Patagonia, Argentina. *Journal of Parasitology*, **77**, 353–6.

Overstreet, R. M. (1997). Parasitological data as monitors of environmental health. *Parassitologia*, **39**, 169–75.

Overstreet, R. M. & Howse, H. D. (1977). Some parasites and diseases of estuarine fish in polluted habitats of Mississippi. *Annals of the New York Academy of Science*, **298**, 427–62.

Owen, R. W. & Pemberton, R. T. (1962). Helminth infections of the starling (*Sturnus vulgaris*) in Northern England. *Proceedings of the Zoological Society of London*, **139**, 557–87.

Papadopoulos, H., Himonas, C., Papazahariadou, M. & Antoniadou-Sotiriadou, K. (1997). Helminths of foxes and other wild carnivores from rural areas in Greece. *Journal of Helminthology*, **71**, 227–31.

Parshad, V. R. & Crompton, D. W. T. (1981). Aspects of acanthocephalan reproduction. *Advances in Parasitology*, **19**, 73–138.

Patrick, M. J. (1991). Distribution of enteric helminths in *Glaucomys volans* L. (Sciuridae): a test for competition. *Ecology*, **72**, 755–8.

Pavlovski, E. N. (1934). Organism as an environment. *Priorda, Moskava*, **1**, 80–91.

Pence, D. B. (1990). Helminth community of mammalian hosts: concepts at the infracomunity, component and compound community levels. In *Helminth Communities: Patterns and Processes*, ed. G. W. Esch, A. O. Bush & J. M. Aho. London & New York: Chapman & Hall, pp. 233–60.

Pence, D. B. & Windberg, L. A. (1984). Population dynamics across selected habitat variables of the helminth community in coyotes, *Canis latrans*, from South Texas. *Journal of Parasitology*, **70**, 735–46.

Pennycuick, L. (1971). Frequency distributions of parasites in a population of three-spined sticklebacks, *Gasterosteus aculeatus* L., with particular reference to the negative binomial distribution. *Parasitology*, **63**, 389–406.

Perez-Ponce de Leon, G. & Choudhury, A. (2005). Biogeography of helminth parasites of freshwater fish in Mexico: the search for patterns and processes. *Journal of Biogeography*, **32**, 645–59.

Petrochenko, V. I. (1956). *Acanthocephala of Domestic and Wild Animals*, Vol. 1. Moscow: Izdatel'stvo Akademii Nauk SSSR.

Petrochenko, V. I. (1958). *Acanthocephala of Domestic and Wild Animals*, Vol. 2. Moscow: Izdatel'stovo Akademii Nauk SSSR.

Pianka, E. R. (1974). *Evolutionary Ecology*. New York: Harper & Row.

Pichelin, S. (1997). *Pomphorhynchus heronensis* sp.nov. (Acanthocephala: Pomphorhynchidae) from *Lutjanus carponotatus* (Lutjanidae) from Heron Island, Australia. *Records of the South Australian Museum*, **30**, 19–27.

Pineda-Lopez, R. (1994). Ecology of the helminth communities of cichlid fish in the flood plains of southeastern Mexico. Unpublished Ph.D. Thesis: University of Exeter.

Pippy, J. H. C. (1969). *Pomphorhynchus laevis* (Zoega) Muller, 1776 (Acanthocephala) in Atlantic salmon (*Salmo salar*) and its use as a biological tag. *Journal of the Fisheries Research Board of Canada*, **26**, 909–19.

Pippy, J. H. C. (1980). The value of parasites as biological tags in Atlantic salmon at West Greenland. *Rapport et procés-verbaux du Conseil International pour l' Exploration de la Mer*, **176**, 76–81.

Polyanski, Y. I. (1961). Ecology of parasites of marine fishes. In *Parasitology of Fishes*, ed. V. A. Dogiel, G. K. Petrushevski & Y. I. Polyanski, Edinburgh & London: Oliver & Boyd, pp. 48–83.

Polyanski, Y. I. (1966). *Parasites of the Fish of the Barents Sea*. Jerusalem: IPST.

Poulin, R. (1992). Determinants of host-specificity in parasites of freshwater fishes. *International Journal for Parasitology*, **22**, 753–8.

Poulin, R. (1994a). Meta-analysis of parasite induced behavioural changes. *Animal Behaviour*, **48**, 137–46.

Poulin, R. (1994b). The evolution of parasite manipulation of host behaviour: a theoretical analysis. *Parasitology*, **109**, S109–S118.

Poulin, R. (1995). 'Adaptive' changes in the behaviour of parasitised animals: a critical review. *International Journal for Parasitology*, **25**, 1371–83.

Poulin, R. (1998). *Evolutionary Ecology of Parasites: From Individuals to Communities*. London: Chapman & Hall.

Poulin, R. (2001). Another look at the richness of helminth communities in tropical freshwater fish. *Journal of Biogeography*, **28**, 737–43.

Poulin, R. (2002). Qualitative and quantitative aspects of recent research on helminth parasites. *Journal of Helminthology*, **76**, 373–6.

Poulin, R. (2003). The decay of similarity with geographical distance in parasite communities of vertebrate hosts. *Journal of Biogeography*, **30**, 1609–15.

Poulin, R. (2005). Relative infection levels and taxonomic distances among the host species used by a parasite: insights into parasite specialisation. *Parasitology*, **130**, 109–15.

Poulin, R. & Morand, S. (1999). Geographical distances and the similarity among parasite communities of conspecific host populations. *Parasitology*, **119**, 369–74.

Poulin, R. & Mouillot, D. (2003). Parasite specialization from a phylogenetic perspective: a new index of host specificity. *Parasitology*, **126**, 473–80.

Poulin, R. & Mouillot, D. (2005a). Host specificity and the probability of discovering species of helminth parasites. *Parasitology*, **130**, 709–15.

Poulin, R. & Mouillot, D. (2005b). Combining phylogenetic and ecological information into a new index of specificity. *Journal of Parasitology*, **91**, 511–14.

Poulton, M. J. & Thompson, D. J. (1987). The effects of the acanthocephalan parasite *Pomphorhynchus laevis* on mate choice in *Gammarus pulex*. *Animal Behaviour*, **33**, 1577–99.

Price, P. W. (1980). *Evolutionary Biology of Parasites*. Princeton: Princeton University Press.

Price, P. W. (1984). Communities of specialists: vacant niches in ecological and evolutionary time. In *Ecological Communities*, ed. D. R. Strong, D. Simberloff, L. G. Abele & A. B. Thistle, Princeton: Princeton University Press, pp. 510–23.

Price, P. W. (1986). Evolution in parasite communities. In *Parasitology–Quo vadit? Proceedings of the Sixth International Congress of Parasitology*, ed. M. J. Howell. Brisbane: Australian Academy of Sciences, pp. 209–14.

Pulgar, J., Aldana, M., Vergara, E. & George-Nascimento, M. (1995). La conducta de la jaiba estuarine *Hemigrapsus crenulatus* (Milne-Edwards 1837) en relacion al parasitismo por el acantocefalo *Profilicollis antarcticus* (Zdzitowiecki 1985). *Revista Chilena de Historia Natural*, **68**, 439–50.

Rankin, J. S. (1937). An ecological study of parasites of some North American salamanders. *Ecological Monographs*, **7**, 169–269.

Rauque, C. A., Semenas, L. G. & Viozzi, G. P. (2002). Post-cyclic transmission in *Acanthocephalus tumescens* (Acanthocephala: Echinorhynchidae). *Folia Parasitologica*, **49**, 127–30.

Rauque, C. A., Viozzi, G. P. & Semenas, L. G. (2003). Component population study of *Acanthocephalus tumescens* (Acanthocephala) from Lake Moreno, Argentina. *Folia Parasitologica*, **50**, 72–8.

Reyda, F. B. & Nickol, B. B. (2001). A comparison of biological performances among a laboratory-isolated population and two wild populations of *Moniliformis moniliformis*. *Journal of Parasitology*, **87**, 330–8.

Roberts, L. S. & Janovy J., Jr. (1996). *Foundations of Parasitology*. 5th edn. Dubuque, IA: Wm. C. Brown.

Roca, V. & Hornero, M. J. (1994). Helminth infracommunities of *Podarcis pitusensis* and *Podarcis lilfordi* (Sauria: Lacertidae) from the Balearic Islands (western Mediterranean basin). *Canadian Journal of Zoology*, **72**, 658–64.

Rohde, K. (1993). *Ecology of Marine Parasites*, 2nd edn. Wallingford. CAB International.

Rohde, K. (1994). Niche restriction in parasites: proximate and ultimate causes. *Parasitology*, **109**, S69–S84.

Rohde, K. (1998). Is there a fixed number of niches for endoparasites of fish? *International Journal for Parasitology*, **28**, 1861–5.

Roubal, F. R. (1993). Comparative histopathology of *Longicollum* (Acanthocephala: Pomphorhynchidae) infection in the alimentary tract and spleen of *Acanthopagrus australis* (Pisces: Sparida). *International Journal for Parasitology*, **23**, 391–7.

Rumpus, A. E. (1973). The parasites of *Gammarus pulex* in the River Avon, Hampshire. Unpublished Ph.D. thesis, University of Exeter.

Rumpus, A. E. & Kennedy, C. R. (1974). The effect of the acanthocephalan *Pomphorhynchus laevis* upon the respiration of its intermediate host *Gammarus pulex*. *Parasitology*, **68**, 271–84.

Salgado-Maldonado, G. & Kennedy, C. R. (1997). Richness and diversity of helminth communities in a tropical cichlid fish in Mexico. *Parasitology*, **114**, 581–90.

Sanchez-Ramirez, C. & Vidal-Martinez, V. M. (2002). Metazoan parasite infracommunities of Florida pompano (*Trachinotus carolinus*) from the coast of the Yucatan peninsula, Mexico. *Journal of Parasitology*, **88**, 1087–94.

Saraiva, A. & Eiras, J. C. (1996). Parasite community of the European eel, *Anguilla anguilla* (L.), in the River Este, Northern Portugal. *Research and Reviews in Parasitology*, **56**, 179–83.

Schabuss, M., Konecny, R., Belpaire, C. & Schiemer, F. (1997). Endoparasitic helminths of the European eel, *Anguilla anguilla*, from four disconnected meanders from the rivers Leie and Scheldt in western Flanders, Belgium. *Folia Parasitologica*, **44**, 12–18.

Schabuss, M., Kennedy, C. R., Konecny, R., Grillitsch, B., Schiemer, F. & Herzig, A. (2005). Long-term investigation of the composition and richness of intestinal helminth communities in the stocked population of eel, *Anguilla anguilla*, in Neusiedler See, Austria. *Parasitology*, **130**, 185–94.

Scheef, G., Sures, B. & Taraschewski, H. (2000). Cadmium accumulation in *Moniliformis moniliformis* (Acanthocephala) from experimentally infected rats. *Parasitology Research*, **86**, 688–91.

Schludermann, C., Konecny, R., Laimgruber, S. *et al.* (2003). Fish macroparasites as indicators of heavy metal pollution in river sites in Austria. *Parasitology*, **126**, S61–S69.

Schmidt, G. D. (1971). Acanthocephalan infections of man, with two new records. *Journal of Parasitology*, **57**, 582–4.

Schmidt, G. D. (1983). What is *Echinorhynchus pomatostomi* Johnston and Cleland, 1912? *Journal of Parasitology*, **69**, 397–9.

Schmidt, G. D. (1985). Development and life cycles. In *Biology of the Acanthocephala*, ed. D. W. T. Crompton & B. B. Nickol, Cambridge: Cambridge University Press, pp. 273–305.

Schmidt, G. D. & Hugghins, E. J. (1973). Acanthocephala of South American fishes: 2. Palaeacanthocephala. *Journal of Parasitology*, **59**, 836–8.

Schmidt, G. D. & Olsen, O. W. (1964). Life cycle and development of *Prosthorhynchus formosus* (Van Cleave, 1918) Travassos, 1926, an acanthocephalan parasite of birds. *Journal of Parasitology*, **50**, 721–30.

Schmidt, G. D. & Paperna, I. (1978). *Sclerocollum rubrimaris* Gen. et Sp. N. (Rhadinorhynchidae: Gorgorhynchinae) and other acanthocephala of marine fishes from Israel. *Journal of Parasitology*, **64**, 846–50.

Schmidt, G. D., Walley, H. D. & Wijek, D. S. (1974). Unusual pathology in a fish due to the acanthocephalan *Acanthocephalus jacksoni* Bullock, 1962. *Journal of Parasitology*, **60**, 730–1.

Segovia, J. M., Torres, J., Miquel, J., Llaneza, L. & Feliu, C. (2001). Helminths in the wolf, *Canis lupus*, from north-western Spain. *Journal of Helminthology*, **75**, 183–92.

Seidenberg, A. J. (1973). Ecology of the acanthocephalan, *Acanthocephalus dirus* (Van Cleave, 1931) in its intermediate host, *Asellus intermedius* Forbes (Crustacea: Arthropoda). *Journal of Parasitology*, **59**, 957–62.

Semenas, L., Ortubay, S. & Ubeda, C. (1992). Studies on the development and life-history of *Pomphorhynchus patagonicus* Ortubay, Ubeda, Semenas and Kennedy, 1991 (Palaeacanthocephala). *Research and Reviews in Parasitology*, **52**, 89–93.

Serov, V. G. (1984). Fecundity of *Acanthocephalus lucii* (Echinorhynchidae). *Parazitologia*, **18**, 280–5.

Sharpilo, V. P., Biserkov, V., Kostadinova, A., Behnke, J. M. & Kuzmin, Y. I. (2001). Helminths of the sand lizard, *Lacerta agilis* (Reptilia, Lacertidae) in the Palaearctic: faunal diversity and spatial patterns of variation in the composition and structure of component communities. *Parasitology*, **123**, 389–400.

Sherwin, F. J. & Schmidt, G. D. (1988). Helminths of swallows of the mountains of Colorado, including *Acuaria coloradebsis* N.Sp. (Nematoda: Spirurata). *Journal of Parasitology*, **74**, 336–8.

Shostak, A. W., Dick, T. A., Szalai, A. J. & Bernier, L. M. J. (1986). Morphological variability in *Echinorhynchus gadi*, *E. leidy* and *E. salmonis* (Acanthocephala: Echinorhynchidae) from fishes in northern Canadian waters. *Canadian Journal of Zoology*, **64**, 985–95.

Siddal, R. & Sures, B. (1998). Uptake of lead by *Pomphorhynchus laevis* cystacanths in *Gammarus pulex* and immature worms in chub (*Leuciscus cephalus*). *Parasitology Research*, **84**, 573–7.

Simberloff, D. (1990). Free-living communities and alimentary tract helminths: hypotheses and pattern analyses. In *Parasite Communities: Patterns and Processes*, ed. G. W. Esch, A. O. Bush & J. M. Aho. London & New York: Chapman & Hall, pp. 289–319.

Smales, L. R. (1988). *Plagiorhynchus (Prosthorhynchus) cylindraceus* (Goeze, 1782) Schmidt & Kuntz, 1966 from the Australian bandicoots, *Paramela gunnii* Gray, 1838, and *Isoodon obesulus* (Shaw, 1797). *Journal of Parasitology*, **74**, 1062–4.

Smith, R. A., Kennedy, M. L. & Wilhelm, W. E. (1985). Helminth parasites of the racoon (*Procyon lotor*) from Tennessee and Kentucky. *Journal of Parasitology*, **71**, 599–603.

Smith Trail, D. R. (1980). Behavioural interactions between parasitised and hosts: host suicide and the evolution of complex life cycles. *American Naturalist*, **118**, 715–25.

Smyth, J. D. (1994). *Introduction to Animal Parasitology*, 3rd edn. Cambridge: Cambridge University Press.

Snyder, D. E. & Fitzgerald, P. R. (1985). Helminth parasites from Illinois racoons (*Procyon lotor*). *Journal of Parasitology*, **71**, 274–8.

Solaymani-Mohammadi, D., Mobedi, I., Rezaian, M. *et al.* (2003). Helminth parasites of the wild boar, *Sus scrofa*, in Luristan province, western Iran and their public health significance. *Journal of Helminthology*, **77**, 263–7.

Sparkes, T. C., Wright, V. M., Renwick, D. T., Weil, K. A., Talkington, J. A. & Milhalyov, M. (2004). Intra-specific host sharing in the manipulative parasite *Acanthocephalus dirus*: does conflict occur over host modification? *Parasitology*, **129**, 335–40.

Sprent, J. F. A. (1992). Parasites lost. *International Journal for Parasitology*, **22**, 139–51.

Starling, J. A. (1985). Feeding, nutrition and metabolism. In *Biology of the Acanthocephala*, ed. D. W. T. Crompton & B. B. Nickol. Cambridge: Cambridge University Press, pp. 125–212.

Stock, T. M. & Holmes, J. C. (1988). Functional relationships and microhabitat distributions of enteric helminths of grebes (Podicipedidae): the evidence for interactive communities. *Journal of Parasitology*, **74**, 214–27.

Styczynska, E. (1958). Acanthocephala of the biocenosis of Druzno Lake (Parasitofauna of the biocenosis of Druzno Lake part 6). *Acta Parasitologica Polonica*, **6**, 195–211.

Sures, B. (2002). Competition for minerals between *Acanthocephalus lucii* and its definitive host perch (*Perca fluviatilis*). *International Journal for Parasitology*, **32**, 1117–22.

Sures, B. (2003). Accumulation of heavy metals by intestinal helminths in fish: an overview and perspective. *Parasitology*, **126**, S53–S60.

Sures, B. (2004). Environmental parasitology: relevancy of parasites in monitoring environmental pollution. *Trends in Parasitology*, **20**, 170–7.

Sures, B. & Reimann, N. (2003). Analysis of trace metals in the Antarctic host-parasite system *Notothenia coriiceps* and *Aspersentis megarhynchus* (Acanthocephala) caught at King George Island, South Shetland Islands. *Polar Biology*, **26**, 680–6.

Sures, B. & Siddal, R. (1999). *Pomphorhynchus laevis*: the intestinal acantho-cephalan as a lead sink for its fish host, chub (*Leuciscus cephalus*). *Experimental Parasitology*, **93**, 66–72.

Sures, B. & Siddal, R. (2003). *Pomphorhynchus laevis* (Palaeoacanthocephala) in the intestine of chub (*Leuciscus cephalus*) as an indicator of metal pollution. *International Journal for Parasitology*, **33**, 65–70.

Sures, B. & Streit, B. (2001). Eel parasite diversity and intermediate host abundance in the River Rhine, Germany. *Parasitology*, **123**, 185–91.

Sures, B. & Taraschewski, H. (1995). Cadmium concentrations in two adult acanthocephalans, *Pomphorhynchus laevis* and *Acanthocephalus lucii*, as compared with their fish hosts and cadmium and lead levels in larvae of *A. lucii* as compared with their crustacean host. *Parasitology Research*, **81**, 494–7.

Sures, B., Taraschewski, H. & Jackwerth, E. (1994). Lead accumulation in *Pomphorhynchus laevis* and its hosts. *Journal of Parasitology*, **80**, 355–7.

Sures, B., Taraschewski, H. & Siddal, R. (1997a). Heavy metal concentrations in adult acanthocephalans and cestodes compared to their fish hosts and to established free-living bioindicators. *Parassitologia*, **39**, 213–18.

Sures, B., Taraschewski, H. & Rydlo, M. (1997b). Intestinal fish parasites as heavy metal indicators: a comparison between *Acanthocephalus lucii* (Palaeoacanthocephala) and the zebra mussesl, *Dreissena polymorpha*. *Bulletin of Environmental Contamination and Toxicology*, **59**, 14–21.

Sures, B., Knopf, K., Wurtz, J. & Hirt, J. (1999a). Richness and diversity of parasite communities in European eels *Anguilla anguilla* of the River Rhine, Germany, with special reference to helminth parasites. *Parasitology*, **119**, 323–30.

Sures, B., Siddal, R. & Taraschewski, H. (1999b). Parasites as accumulation indicators of heavy metal pollution. *Parasitology To-day*, **15**, 16–21.

Sures, B., Franken, M. & Taraschewski, H. (2000a). Element concentrations in the archiacanthocephalan *Macracanthorhynchus hirudinaceus* compared with those in the porcine host from a slaughterhouse in La Paz, Bolivia. *International Journal for Parasitology*, **30**, 1071–6.

Sures, B., Jurges, G. & Taraschewski, H. (2000b). Accumulation and distribution of lead in the acanthocephalan *Moniliformis moniliformis* from experimentally infected rats. *Parasitology*, **121**, 427–33.

Sures, B., Zimmerman, S., Sonntag, C., Stuben, D. & Taraschewski, H. (2003). The acanthocephalan *Paratenuisentis ambiguus* as a sensitive indicator of the precious metals Pt and Rh from automobile catalytic converters. *Environmental Pollution*, **122**, 401–5.

Szalai, A. L. & Dick, T. A. (1987). Intestinal pathology and site specificity of the acanthocephalan *Neoechinorhynchus carpiodi* Dechtair, 1968, in quillback *Carpiodes cyprinus* (Lesueur). *Journal of Parasitology*, **73**, 467–75.

Taraschewski, H. (2000). Host-parasite interactions in Acanthocephala: a morphological approach. *Advances in Parasitology*, **46**, 1–179.

Taraschewski, H., Moravec, F., Lamah, T. & Anders, K. (1987). Distribution and morphology of two helminths recently introduced into European eel populations: *Anguillicola crassus* (Nematoda, Dracunculoidea) and *Paratenuisentis ambiguus* (Acanthocephala; Tenuisentidae). *Diseases of Aquatic Organisms*, **3**, 167–76.

Tedla, S. & Fernando, C. H. (1970). Some remarks on the ecology of *Echinorhynchus salmonis* (Muller, 1784). *Canadian Journal of Zoology*, **48**, 317–21.

Thielen, F., Zimmerman, S., Baska, F., Taraschewski, H. & Sures, B. (2004). The intestinal parasite *Pomphorhynchus laevis* (Acanthocephala) from barbel as a bioindicator for metal pollution in the Danube River near Budapest, Hungary. *Environmental Pollution*, **129**, 421–9.

Thomas, F., Mete, K., Helluy, S. *et al.* (1997). Hitch-hiker parasites or how to benefit from the strategy of another parasite. *Evolution*, **51**, 1316–18.

Thomas, F., Renaud, F. & Poulin, R. (1998). Exploitation of manipulators: 'hitch-hiking' as a parasite transmission strategy. *Animal Behaviour*, **56**, 199–206.

Thompson, A. B. (1985a). Analysis of *Profilicollis botulus* (Acanthocephala: Echinorhynchidae) burdens in the shore crab, *Carcinus maenas*. *Journal of Animal Ecology*, **54**, 595–604.

Thompson, A. B (1985b). Transmission dynamics of *Profilicollis botulus* (Acanthocephala) from crabs (*Carcinus maenas*) to eider ducks (*Somateria mollisima*) on the Ythan estuary, N.E. Scotland. *Journal of Animal Ecology*, **54**, 605–16.

Thompson, A. B. (1985c). *Profilicollis botulus* (Acanthocephala) abundance in the eider duck (*Somateria mollisima*) on the Ythan estuary, Aberdeenshire. *Parasitology*, **91**, 563–75.

Threlfall, W. (1967). Studies on the helminth parasites of the herring gull, *Larus argentatus* Pontop., in northern Caernarvonshire and Anglesey. *Parasitology*, **57**, 431–53.

Toft, C. A. (1991). An ecological perspective: the population and community consquences of parasitism. In *Parasite-Host Associations: Coexistence or Conflict?* ed. C. A. Toft, A. Aeschlimann & L. Bolis. Oxford: Oxford University Press, pp. 319–43.

Tokeson, J. P. E. & Holmes, J. C. (1982). The effects of temperature and oxygen on the development of *Polymorphus marilis* (Acanthocephala) in *Gammarus lacustris* (Amphipoda). *Journal of Parasitology*, **68**, 112–19.

Torchin, M. E., Lafferty, K. D. & Kuris, A. M. (2002). Parasites and marine invasions. *Parasitology*, **124**, S137–S151.

Torchin, M. E., Lafferty, K. D., Dobson, A. P., McKenzie, V. J. & Kuris, A. M. (2003). Introduced species and their missing parasites. *Nature, London*, **421**, 628–30.

Torres, J., Garcia-Perea, R., Gisbert, J. & Feliu, C. (1998). Helminth fauna of the Iberian lynx, *Lynx pardinus*. *Journal of Helminthology*, **72**, 221–6.

Torres, J., Feliu, C., Fernandez-Moran, J. *et al.* (2004). Helminth parasites of the Eurasian otter *Lutra lutra* in southwest Europe. *Journal of Helminthology*, **78**, 353–9.

Townsend, C. R., Harper, J. L. & Begon, M. (2000). *Essentials of Ecology*. Malden, Massachusetts: Blackwell Science.

Trejo, A. (1992). A comparative study of the host-parasite relationships of *Pomphorhynchus patagonicus* (Acanthocephala) in two species of fish from Lake Rosario (Chubut, Argentina). *Journal of Parasitology*, **78**, 711–15.

Trejo, A. (1994). Observations of the host specificity of *Pomphorhynchus patagonicus* (Acanthocephala) from the Alicura Reservoir (Patagonia, Argentina). *Journal of Parasitology*, **80**, 829–30.

Ubeda, C., Trejo, A., Semenas, L. & Ortubay, S. (1994). Status of three different fish hosts of *Pomphorhynchus patagonicus* Ortubay, Ubeda, Semenas et Kennedy, 1991 (Acanthocephala) in Lake Rosario (Argentina). *Research and Reviews in Parasitology*, **54**, 87–92.

Uglem, G. L. (1972). The life cycle of *Neoechinorhynchus cristatus* Lynch, 1936 (Acanthocephala) with notes on the hatching of eggs. *Journal of Parasitology*, **58**, 911–20.

Uglem, G. L. & Beck, S. M. (1972). Habitat specificity and correlated aminopeptidase activity in the acanthocephalans *Neoechinorhynchus cristatus* and *N. crassus*. *Journal of Parasitology*, **58**, 911–20.

Uznanski, R. L. & Nickol, B. B. (1970). Structure and function of the fibrillar coat of *Leptorhynchoides thecatus* eggs. *Journal of Parasitology*, **62**, 569–73.

Uznanski, R. L. & Nickol, B. B. (1980). Parasite population regulation: lethal and sublethal effects of *Leptorhynchoides thecatus* (Acanthocephala; Rhadinorhynchidae) in *Hyalella azteca* (Amphipoda). *Journal of Parasitology*, **66**, 121–6.

Uznanski, R. L. & Nickol, B. B. (1982). Site selection, growth, and survival of *Leptorhynchoides thecatus* (Acanthocephala) during the prepatent period in *Lepomis cyanellus*. *Journal of Parasitology*, **68**, 686–90.

Vainola, R., Valtonen, E. T. & Gibson, D. I. (1994). Molecular systematics in the acanthocephalan genus *Echinorhynchus* (*sensu lato*) in northern Europe. *Parasitology*, **108**, 105–14.

Valtonen, E. T. (1980). *Metechinorhynchus salmonis* (Muller, 1780) (Acanthocephala) as a parasite of the whitefish in the β othnian Bay: I. Seasonal relationships between infection and fish size. *Acta Parasitologica Polonica*, **27**, 293–300.

Valtonen, E. T. (1983). On the ecology of *Echinorhynchus salmonis* and two *Corynosoma* species (Acanthocephala) in the fish and seals. Acta Universitatis Ouluensis Series A, No. **156**, *Biologica*, **22**, 1–49.

Valtonen, E. T. & Crompton, D. W. T. (1990). Acanthocephala in fish from the Bothnian Bay, Finland. *Journal of Zoology*, **220**, 619–39.

Valtonen, E. T. & Helle, E. (1988). Host-parasite relationships between two seal populations and two species of *Corynosoma* (Acanthocephala) in Finland. *Journal of Zoology*, **214**, 361–71.

Valtonen, E. T., Holmes, J. C. & Koskivaara, M. (1997). Eutrophication, pollution and fragmentation: effects on the parasite communities in roach and perch in four lakes in Finland. *Parassitologia*, **39**, 233–6.

Valtonen, E. T., Holmes, J. C., Aronen, J. & Rautalahti, I. (2003). Parasite communities as indicators of recovery from pollution: parasites of roach (*Rutilus rutilus*) and perch (*Perca fluviatilis*) in Central Finland. *Parasitology*, **126**, S43–S52.

Valtonen, E. T., Helle, E. & Poulin, R. (2004). Stability of *Corynosoma* populations with fluctuating population densities of the seal definitive host. *Parasitology*, **129**, 635–42.

Van Maren, M. J. (1979a). Structure and dynamics of the French upper Rhone ecosystems: XII. An inventory of helminth fish parasites from the upper Rhone river (France). *Bulletin Zoologisch Museum, Universiteit van Amsterdam*, **6**, 189–99.

Van Maren, M. J. (1979b). The amphipod *Gammarus fossarum* Koch (Crustacea) as intermediate host for some helminth parasites, with notes on their occurrence in the final host. *Bijdragen tot de Dierkunde*, **48**, 97–110.

Vidal-Martinez, V. M. & Kennedy, C. R. (2000a). Zoogeographical determinants of the composition of the helminth fauna of neotropical cichlid fish. In. *Metazoan Parasites in the Neotropics: A Systematic and Ecological Perspective*, ed. G. Salgado-Maldonado, A. N. Garcia-Aldrete & V. M. Vidal-Martinez. Mexico: Instituto de Biologia, UNAM, pp. 227–90.

Vidal-Martinez, V. M. & Kennedy, C. R. (2000b). Potential interactions between the intestinal helminths of the cichlid fish *Cichlosoma synspilum* from southeastern Mexico. *Journal of Parasitology*, **86**, 691–5.

Vidal-Martinez, V. M. & Poulin, R. (2003). Spatial and temporal repeatability in parasite community structure of tropical fish hosts. *Parasitology*, **127**, 387−98.

Vidal-Martinez, V. M., Aguirre-Macedo, M. L., Scholz, T., Gonzales-Solis, D. & Mendoza-Franco, E. F. (2001). *Atlas of the Helminth Parasites of Cichlid Fish of Mexico*. Prague: Academia.

Vidal-Martinez, V. M., Aguirre-Macedo, M. L., Norena-Barroso, E., Gold-Bouchot, G. & Caballero-Pinzon, P. I. (2003). Potential interactions between metazoan parasites of the Mayan catfish *Ariopsis assimilis* and chemical pollution in Chetumal Bay, Mexico. *Journal of Helminthology*, **77**, 173−84.

Vrcibradic, D., Rocha, C. F. D., Bursey, C. R. & Vicente, J. J. (2002). Helminth communities of two sympatric skinks (*Mabuya agilis* and *Mabuya macrorhyncha*) from two 'restinga' habitats in southeastern Brazil. *Journal of Helminthology*, **76**, 355−61.

Walkey, M. (1967). The ecology of *Neoechinorhynchus rutili* (Muller). *Journal of Parasitology*, **53**, 795−804.

Whitfield, P. J. (1970). The egg sorting function of the uterine bell of *Polymorphus minutus* (Acanthocephala). *Parasitology*, **61**, 671−82.

Whitfield, P. J. (1979). *The Biology of Parasitism*. London: Edward Arnold.

Williams, H. & Jones, A. (1994). *Parasitic Worms of Fish*. London: Taylor & Francis.

Williams, H. H. & MacKenzie, K. (2003). Marine parasites as pollution indicators: an update. *Parasitology*, **126**, S27−S41.

Williams, H. H., MacKenzie, K. & McCarthy, A. M. (1992). Parasites as biological indicators of the population biology, migrations, diet and phylogenetics of fish. *Reviews in Fish Biology and Fisheries*, **2**, 144−76.

Willingham, A. L., Ockens, N. W., Kapel, C. M. O. & Monrad, J. (1996). A helminthological survey of wild red foxes (*Vulpes vulpes*) from the metropolitan area of Copenhagen. *Journal of Helminthology*, **70**, 259−63.

Wongkham, W. & Whitfield, P. J. (2004). *Pallisentis rexus* from the Chiang Mai Basin, Thailand: ultrastructural studies on egg envelope development and the mechanism of egg expansion. *Journal of Helminthology*, **78**, 77−85.

Yamaguti, S. (1963). *Acanthocephala*. Vol. V of *Systema Helminthum*. New York: John Wiley.

Yang, T. & Liao, X. (2001). Seasonal population dynamics of *Neoechinorhynchus qinghaiensis* in the carp, *Gymnocypris przewalskii przewalskii*, from Qinghai Lake, China. *Journal of Helminthology*, **75**, 93−8.

Zdzitowiecki, K. (2001). Acanthocephala occurring in intermediate hosts, amphipods, in Admiralty Bay (South Shetland Islands, Antarctica). *Acta Parasitologica*, **46**, 202−7.

Zimmermann, S., von Bohlen, A., Nesserschmidt, J. & Sures, B. (2005). Accumulation of the precious metals platinum, palladium and rhodium from automobile catalytic converters in *Paratenuisentis ambiguus* as compared with its fish host, *Anguilla anguilla*. *Journal of Helminthology*, **79**, 85−9.

Index

Printed in the United States
by Baker & Taylor Publisher Services